Also by Isaac Asimov

FOUNDATION
FOUNDATION AND EMPIRE
FACT AND FANCY
LIFE AND ENERGY
SECOND FOUNDATION
THE UNIVERSE

FACT AND FANCY

Isaac Asimov

 DISCUS BOOKS/ PUBLISHED BY AVON

AVON BOOKS
A division of
The Hearst Corporation
959 Eighth Avenue
New York, New York 10019

First Printing (Discus Edition), March, 1972

Printed in the U.S.A.

To the fine gentlemen responsible:

Joseph W. Ferman
Robert P. Mills

Introduction

The dullness of fact is the mother of fiction. How many lies originate out of no desire to escape punishment, amass undue credit, or gain an end, but simply to make a good story. And a story, endlessly repeated, grows by accretion of false detail, so that the fish one almost catches becomes larger, the retort to the boss more biting, the fright more frightful, and the narrow escape incredibly hairlike.

Fortunate the man who, by profession, can lie freely and call it a novel. If he lies skillfully and evocatively enough, revealing, as he does so, a piece of mankind to itself, he may even attain immortality and the eternal gratitude of mankind in place of the impatient sneer that is the usual reward of the liar.

And contrariwise, sad the lot of the man who, in his writing, finds himself in a field so dedicated to fact, however dull, that the least deviation in a moment of carelessness is viewed with serious alarm.

To what field can I be referring but that of Science, the cold and rigid apostle of "truth-as-we-now-see-it." The facts, gentlemen, and nothing but the facts, for careful eyes are narrowly watching.

I call you, then, to witness my own peculiarly exacerbated difficulty as a science writer, for I began my writing career in fiction and in twenty years wrote well over a hundred short stories and novelettes plus a dozen or more novels. My instinct for embroidery is so well-developed now, one might say hypertrophied, that it quivers in agony at the first distant clumping that heralds the approach of a dull fact.

Surely there must be some middle ground between the four-squareness of the fact, the solidity of its flat feet, the thud of its stone sandals, and the iridescent gauziness of a complete lie flitting its way through the ether.

I am told, and I *know*, that Science is fascinating and ad-

7

venturesome; that it bears the burning mark of the greatest of all frontiers, that of the human mind facing the darkly infinite sea of the unknown that hems it in.

To me there is no pleasure in the writing of science if I cannot make the effort of capturing some of the gauze and iridescence that belong to the truths snatched out of the chaos of ignorance, far more than they can ever belong to some puny lie.

I can think of no better word for that gauze of truth than "fancy."

The distance of the moon from the earth and from the sun: the sizes and motions of the three bodies are facts. To deduce therefrom the vision of an eclipse of the sun by the earth, as seen from the moon, is no lie, even though no such sight has yet been seen by the corporeal eye of a living man. The very fact that it is basic truth, not yet fully revealed, makes it far more fascinating than any lie could be. It is fancy.

The sun has a family of nine known major planets. That is fact. A tenth planet may exist, for all we know, and if so, certain facts may be deduced from what we already know about the rest of the system. That is fancy.

There are—possibly—icy asteroids circling the sun, so far out in space, that the whole solar system we know today shrinks almost to a dot in comparison. It is possible that matter comes into existence at a terribly slow rate and disappears again just as slowly—well, perhaps. Man has seen distant stars explode and grow incredibly bright. He has never seen a neighboring star do so and grow in brightness to rival our own sun for a few weeks. The vision of that and an infinity of other wonders belong to fancy. The art of the lie can touch nothing so grand.

And so I resolve my dilemma. The hospitable pages of the *Magazine of Fantasy and Science Fiction* are thrown open each month to me, and there, under the kindly egis of editor Robert P. Mills* and free of all limitations and censorship, I fit out my facts, as best as ever I can, with the gauzy wings of fancy, and send them flying.

The following collection of articles are (with one exception) from the pages of that magazine, and if there is even half the fun in the reading that there was in the writing, I will be delighted.

---

* Now (1963) a full-time, but still kind, literary agent; the magazine's editorship has passed to Avram Davidson, who has been equally hospitable to the articles succeeding the ones in this book. I. A.

PART I / THE EARTH
AND AWAY

1 Life's Bottleneck

Villains on a cosmic scale are where you find them, and the imagination has found some majestic ones indeed, including exploding suns and invading Martians. Real life, in recent years, has found some actual villains that would have seemed most imaginary not too long ago, as, for instance, nuclear bombs and melting icecaps.

But there are always a few more, if you look long enough, —like sanitary plumbing and modern sewage disposal.

Well, let me explain.

To begin with, let's consider the ocean, the mother of all things living. Out of its substance, some billions of years ago, life formed, utilizing for the purpose the various types of atoms found in the ocean, though it had to juggle the proportions a bit.

For instance, the ocean is mainly water, and so is living tissue. The ocean is 97 per cent water by weight, while living things in the ocean are about 80 per cent water, generally speaking.

However, this is not quite a fair comparison. The water molecule is made up of two hydrogen atoms and an oxygen atom. In the ocean, water itself is the only substance, to speak of, which contains these atoms. In living matter, however, hydrogen and oxygen are contained in many of the constituent molecules other than water; and all this hydrogen and oxygen came from water originally. This "hydrogen-and-oxygen-elsewhere" should be counted with water, therefore.

To get a more proper panorama, let's consider the percentage by weight of each constituent type of atom. We can do this for the ocean, and for the copepod, a tiny crustacean which is one of the more common forms of the ocean's swarming animal life. The results are given in Table 1.

The column headed "Concentration Factor" in that table is the most important part of it. It represents the ratio of the percentage of a particular substance in living tissue to the percentage of the same substance in the environment.

TABLE 1

	Per Cent Composition of Ocean	Per Cent Composition of Copepod	Concentration Factor
Oxygen	85.89	79.99	0.93
Hydrogen	10.82	10.21	0.94
Everything else	3.29	9.80	3.35

For instance, oxygen and hydrogen are found in smaller percentage in tissue than in ocean so the concentration factor for each is less than 1, as shown in the table. To convert 100 pounds of ocean (containing 96.71 pounds of hydrogen and oxygen) into 100 pounds of copepod (containing 90.20 pounds of hydrogen and oxygen) 6.51 pounds of hydrogen and oxygen must be gotten rid of.

Whenever the concentration factor for any substance is less than 1, it means that that particular substance, potentially at least, can never be a limiting factor in the multiplication of living things. Life's problem will always be to get rid of it, rather than to collect it.

The situation is the reverse as far as "everything else" is concerned. Here 100 pounds of copepod contains 9.80 pounds of "everything else" while 100 pounds of ocean—out of which the copepod is formed—contains only 3.29 pounds. It takes 335 pounds of ocean to .contain the 9.80 pounds of "everything else."

A concentration factor greater than 1 sets up the possibility of a bottleneck. Ideally, life could multiply in the ocean till the entire ocean had been converted into living tissue. After all, what is there to stop the endless and unlimited multiplication of life?

Well, suppose you begin with 335 pounds of ocean. By the time copepods have multiplied to a total weight of 100 pounds, they have incorporated all the "everything else" in the supply of ocean into their own bodies. There is still 235 pounds of ocean left, but it is pure water and cannot be converted into copepod.

The greater the concentration factor, the more quickly that limit would have been reached and the smaller the fraction of the total environment that can be converted into living tissue.

Of course, I have deliberately simplified the matter, to begin with, in order to make the point. Actually, the "everything else" is a conglomerate of a dozen or so elements, each of which is essential to life, and none of which can be dispensed with.

Each essential element is present in different amounts in the ocean; each is present in different amounts in living tissue. Each, therefore, has its own concentration factor. As soon as any one of them is completely used up, the possibility of the further expansion of life, generally, halts. One form of life can grow at the expense of another, but the total quantity of protoplasm can increase no further.

The essential element with the highest concentration factor is the one first used up and is, therefore, life's bottleneck.

Let's therefore compare the ocean and the copepod in finer detail, omitting the hydrogen and oxygen and just considering the "everything else." This is done in Table 2.

You can see that concentration factors do indeed vary widely from element to element. Only four elements have factors that are really extreme; that is, over a thousand. Of these four, the values for carbon and nitrogen are not really as extreme as they seem, however, since the ocean is not the only source of these elements. There is, for instance, some carbon dioxide in the air, and all of that is available to ocean life.

TABLE 2

	Per Cent Composition of Ocean	Per Cent Composition of Copepod	Concentration Factor
Carbon	0.0031	6.10	2000.
Nitrogen	0.00008	1.52	19,000.
Chlorine	2.04	1.05	0.52
Sodium	1.09	0.54	0.50
Potassium	0.042	0.29	6.9
Sulfur	0.097	0.14	1.4
Phosphorus	0.000011	0.13	12,000.
Calcium	0.0024	0.04	16.5
Magnesium	0.13	0.03	0.23

Iron	0.000002	0.007	3500.
Silicon	0.0004	0.007	17.
Bromine	0.0072	0.0009	0.12
Iodine	0.000005	0.0002	40.

(And the supply of atmospheric carbon dioxide is increasing these days as we burn coal and oil.)

There is also a vast quantity of nitrogen in the air; much more than there is in the ocean. This is available to ocean life, too, at least indirectly, through the activity of nitrogen-fixing bacteria. These convert gaseous nitrogen, which is itself unusable to higher forms of life, into nitrates, which are usable.

For these reasons, neither carbon nor nitrogen can ever be considered bottlenecks against the additional formation of total protoplasm. There is only a finite quantity of both, but long before life feels the pinch in either carbon or nitrogen, there is the shortage of iron and phosphorus to be continued.

And here phosphorus is four times as critical as iron. The copepod, of course, is only one type of life, but in general this pattern carries through. Phosphorus has the highest concentration factor; it is the first element to be used up. Life can multiply until all the phosphorus is gone and then there is an inexorable halt which nothing can prevent.

Even that much is only possible under favorable energy conditions. For it takes energy to concentrate the phosphorus and iron of the ocean to the levels required by living tissue. For that matter, it takes energy to expel enough of the chlorine, sodium, magnesium, and bromine to bring their concentrations down to levels tolerated by living tissue. It also takes energy to convert the simple low-energy compounds of the ocean (even after appropriate concentration or thinning out) into the complicated high-energy compounds characteristic of living tissue.

The energy required is supplied by sunlight, which is inexhaustible in those places where it exists. Where it does exist, plant cells multiply and convert, by photosynthesis, the radiant energy of the sun into the chemical energy of carbohydrates, fats, and proteins. Animals (a form of life making up only a small portion of the total) obtain their energy by eating the plants' cells and metabolizing their tissue substance for the chemical energy it contains.

But sunlight only exists in the top 150 meters of the ocean. Below that, sunlight does not penetrate and plant cells do not grow. It is only in the top 150 meters (the "euphotic zone," from Greek words meaning "good light") that the energy supply itself is not a bottleneck and life can multiply in all its forms until all the phosphorus is used up.

And it does exactly that.

The inorganic phosphorus content of the surface ocean water itself is virtually zero. Just about all the phosphorus it contains is organic; that is, it is found either in the living cells or in the wastes and dead residues thereof.

What happens, then, in the euphotic zone, is a standoff. Animal life eats plant life, while plant life, using animal's wastes as a phosphorus source, grows to replace that portion of itself that has been eaten. The total volume of life is at a steady maximum.

Life below the euphotic zone depends for its existence on an organic rain from above. Animal organisms can swim downward out of the euphotic zone (and plant cells can be forced down by an unlucky current) and there they may be eaten by creatures that live in the sub-euphotic zone regularly.

Also, dead remnants of life drift downward. They are gobbled up by the animal life of the depths (no plant life below the euphotic zone) which in turn contribute, after death, to a continuous drizzle that moves always lower down. In the long run, this perpetually renewed drizzle supports all life down to the very floor of the abyss.

Below the euphotic zone, it is energy, not phosphorus, that is the bottleneck; energy in the form of the organic compounds of this drizzle, which animals can feed on (in addition to each other, of course) and which they can metabolize for energy. Below the euphotic zone, then, there is less life than is necessary to incorporate all the phosphorus of the surroundings. There are, therefore, inorganic phosphorus compounds (phosphates) remaining in the deep ocean water itself.

The organic drizzle represents a loss of phosphorus to the euphotic zone, for dead tissue and animal wastes are rich in that element. If there were nothing to counteract this transfer of phosphorus from the euphotic zone to the depths, the quantity of life in the euphotic zone would inexorably decline along with the concentration of phosphorus and eventually blink out.

Fortunately, there is circulation between the depths and the surface of the ocean. There is an upwelling of water from the abyss, rich in phosphorus, which replaces the phosphorus lost in the organic drizzle. This upwelling is greatest in cold waters such as those of the Antarctic and the North Atlantic. There the chill and heavy surface waters sink and are replaced from the depths. There, consequently, the euphotic zone is richest in phosphorus and can support the greatest concentration of life. (Giant whales, which require a great deal of food for maintenance, for that reason congregate in the Antarctic and North Atlantic. No fools they.)

On the other hand, the warm and light surface waters of the warm areas of Earth remain tenaciously afloat and are not directly replaced by the colder and heavier waters from the depths. They must depend on surface currents from the cold North and South to replenish their phosphorus content. This second-hand supply of phosphorus has already been plundered by the life-forms that reached it first, so the ocean life of the tropics is less rich than that of the colder zones. In warm, land-locked portions of the ocean, such as the Mediterranean Sea, which are relatively sheltered from the phosphorus-relief of even the cold surface current, the ocean life is still less rich.

On the whole, though, there is a balance everywhere in the oceans, and, again on the whole, it is the concentration of phosphorus, life's bottleneck, that dictates the nature of the balance.

The situation with respect to land-based life has special points of interest. Land life is a latecomer to the scene and is still, quantitatively speaking, a minor offshoot of the ocean. Something like 85 per cent of all living matter lives in water; only 15 per cent on land. It is only the fact that *Homo sapiens* happens to live on land that makes us give terrestrial environment the undue attention it receives.

On land, as you would expect of life-forms that had evolved in water, the real bottleneck is the water itself, which no longer surrounds and permeates the life-forms. Land life has cut down on its use of hydrogen and oxygen in consequence. Whereas hydrogen and oxygen together make up about 90 per cent of a copepod, it makes up only about 86 per cent of a land plant like alfalfa, and only 72 per cent of a land animal such as man.

The cut-down is not remarkably extensive, however, and if a territory receives insufficient water, life-forms are sparse regardless of what elements the soil might contain.

Granting the needed water, bottlenecks must next be

sought for among elements other than oxygen or hydrogen. Carbon and nitrogen are eliminated on land for the same reason they were eliminated in the ocean. The atmospheric supply of nitrogen, thanks to nitrogen-fixing bacteria, is ample, and the carbon supply is fleshed out by the atmospheric carbon dioxide.

That leaves elements other than hydrogen, oxygen, carbon, and nitrogen. Leaving those four out, the remaining elements must all be derived from the minerals of the soil, ultimately. For these we can set up Table 3, comparing the percentage composition of Earth's crust with that of an example of terrestrial plant life, such as alfalfa. (On land, as in the sea, plant life predominates quantitatively, and animal life is absolutely dependent upon it. Whatever element is life's bottleneck for plants is therefore the bottleneck for animals as well.)

In some respects, the concentration factors in Table 3 are not as good as they appear. Comparing them with those in Table 2 would make it seem that by and large, soil is so much more concentrated with the various essential elements than is the ocean that life on land ought to outstrip life in the ocean by far.

TABLE 3

	Per Cent Composition of Soil	Per Cent Composition of Alfalfa	Concentration Factor
Phosphorus	0.12	0.706	5.9
Calcium	3.63	0.58	0.16
Potassium	2.59	0.17	0.066
Sulfur	0.052	0.104	2.0
Magnesium	2.09	0.082	0.039
Chlorine	0.048	0.070	1.5
Iron	5.00	0.0027	0.0005
Boron	0.0010	0.0007	0.70
Manganese	0.10	0.00036	0.0036
Zinc	0.0080	0.00035	0.044
Copper	0.0070	0.00025	0.036
Molybdenum	0.00023	0.00010	0.43
Iodine	0.00003	0.0000025	0.08
Cobalt	0.0040	0.0000010	0.00025

Nevertheless, the fact is that elements contained in solid minerals are useless and unavailable to plant life, and consequently, in the long run, to animal life as well. The plant lives on the substances it can extract from solution in the water contained in the soil.

Since the minerals of the soil are relatively insoluble, the watery solution is a thin one indeed and concentration factors are actually very high. That is one reason why land-based life is actually sparser than sea life despite the greater apparent concentration of minerals on land than in the sea.

Furthermore, the material in the soil is not spread evenly. One region may have adequate quantities, let us say, of zinc or copper because of some local deposit, while a neighboring region may be deficient on both and another neighboring region may have a poisonous excess of both. Any element can represent a local bottleneck to life. This is one reason that, even given plenty of sun and rain, one section of land may be less fertile than another.

To be sure, there is an extremely slow soil-homogenizing factor in the land erosion that goes on over the ages, bringing materials from mountaintops to valleys; in the buckling of strata and the scraping of glaciers and the upraising of mountains. In the very long run, then, local deficiencies and excesses don't matter. It is the over-all concentration factor that matters, and there, on land as on the sea, phosphorus is the bottleneck.

Man can take a hand, of course. He can, to the limits imposed by his technology, make up for deficiencies without waiting for geologic processes. He can transfer water from points of excess (with the ocean as the basic source) to points of deficiency. He can do the same for nitrogen (with the air as the basic source) or for calcium or phosphorus.

In doing this, man is, in a way, trying to homogenize the soil and make it more evenly fertile. He is not raising the maximum potential of fertility. The maximum mass of protoplasm which the land can support, like the maximum that the sea can support, is dictated by the phosphorus content. Phosphorus, on both land and sea, has the highest concentration factor; on both land and sea it is life's bottleneck.

Just as there is a standoff in the euphotic zone, so there is a standoff on land. The rain comes down, dissolves tiny quantities of soil, and on this solution, plants grow until all the phosphorus they can grab has been incorporated into their substance. Animals eat the plants and, in the process of living, excrete phosphorus-containing wastes upon which plant life can feed, grow, and replace the amount of itself which animals have eaten.

And just as there is a drizzle out of the euphotic zone of the ocean, so there is a drizzle out of the land. Some of the dissolved materials in the soil inevitably escape the waiting rootlets and are carried by the seeping soil water to brooks and rivers and eventually to the sea.

Any one river in any one second doesn't transfer much in the way of dissolved substances from land to ocean, but all the rivers together pour 9000 cubic miles of water into the oceans each year, and in that amount of water, even a very thin solution amounts to a lot of dissolved material.

The loss of phosphorus, since that is life's bottleneck, is most serious, and it is estimated that 3,500,000 tons of phosphorus are washed from the land into the sea by the rivers each year. Since phosphorus makes up roughly 1 per cent of living matter, that means that the potential maximum amount of land-based protoplasm decreases each year by 350,000,000 tons.

Of course, there may be some device of retransfer from sea to land, just as in the case of the ocean there is a retransfer of phosphorus from the depths to the surface.

One type of retransfer of phosphorus from sea to land involves bird droppings. Some sea birds live on fish and nest on land. Their droppings are rich in phosphorus (derived from fish, which get it from the ocean) and these sea-derived droppings cover their nesting grounds by the ton. This material, called guano, is a valuable commodity since it is an excellent fertilizer just because of its phosphorus content.

However, the phosphorus returned to land in this fashion represents only 3 per cent or less of that washed out to sea. The rest is not returned!

Nor does the phosphorus washed into the sea remain dissolved there. If it did, life in the sea would gradually multiply as life on land diminished, and the total protoplasmic mass on Earth would remain unchanged. Unfortunately the ocean is already holding all it can of the largely insoluble phosphates. New phosphorus washed into the sea simply precipitates as sediment at the sea bottom.

Of course, over geologic periods, the uplifting of sea bottoms exposes new phosphorus-rich soil to start cycles of land fertility over again. At the present moment, though, this long-range view won't help us. With an increasing population, we need increased fertility of the soil just to stay even, and steadily decreasing fertility could spell disaster.

Especially when man is deliberately accelerating the rate at which phosphorus is being lost to the sea.

This is where the new villain comes in. In all advanced regions of Earth (and more and more regions are becoming advanced) internal plumbing is coming into fashion. Elaborate sewer pipes lace cities and phosphorus-rich wastes are carefully and thoroughly flushed into the ocean.

And so soil fertility declines even faster and cannot be replaced by the chemical industry, since more and more of the most necessary chemical, phosphorus, will be at the ocean's bottom, where man himself helped put it and from where man has no way of retrieving it as yet.

Naturally, I am not suggesting that we abandon plumbing and sewers. I am used to sanitation myself and have no real affection for things like typhoid fever and cholera which go along with lack of it.

I am suggesting, though, that while we try to cope with the inevitable disappearance of coal, oil, wood, space between people, and other things that are vanishing as population and per capita power requirements mount recklessly each year, we had better add the problem of disappearing phosphorus to the list, and do what we can to encourage sewage disposal units which process it as fertilizer rather than dump it as waste—or to mine the ocean-floor.

We may be able to substitute nuclear power for coal power, and plastics for wood, and yeast for meat, and friendliness for isolation—but for phosphorus there is neither substitute nor replacement.

2 No More Ice Ages?

We all know that the radioactive ash resulting from the activities of nuclear power plants is dangerous and its disposal a problem to be brooded over. How different from those nice, decent, non-radioactive, old-fashioned coal-burning (or oil-burning) power plants. We can easily put ourselves into the position of the gentleman of the twenty-fifth century moaning thus for the good old days.

Except that the gentleman of the twenty-fifth century may well be sitting there cursing the good old days as he pushes his air conditioner up a notch and wishes that nuclear reactors—radioactive ash and all—had taken over a few centuries sooner than they did.

For coal and oil release an ash also and that ash is also puffed into the atmosphere. The ash of coal and oil isn't radioactive to be sure; it is only good harmless carbon dioxide, which is already present in the atmosphere anyway.

It is only a minor constituent of the atmosphere, 0.04 per cent by weight, but this comes out into big numbers if all the air Earth has is lumped into the scale. The weight of our atmosphere is 5.70×10^{15} tons, so the weight of the carbon dioxide in our atmosphere is 2.28×10^{12} (about two and a quarter trillion) tons.

That carbon dioxide, however, is subjected to some tremendous pushes and pulls. For instance, all plant life depends for existence on the consumption of atmospheric carbon dioxide. Using the energy of sunlight plus hydrogen atoms (obtained from water molecules), the plants convert the carbon dioxide to carbohydrate and then to all the other organic molecules necessary for the plant structure and chemistry.

Lump all the plant life by land and sea (especially the sea

where the algae use up eight times as much carbon dioxide as all land plants put together) and a considerable amount of the gas is used up. Estimates for the carbon dioxide used up by plant life in one year vary from 60 to 200 billion tons. Even allowing the lower figure, it would seem that the carbon dioxide supply of the atmosphere would be used up in about thirty-six years. The larger figure will consume it in less than twelve years. Then all life comes to an end?

A fearful prospect, except that, of course, when an individual plant dies, bacteria attack the tissues and convert the carbon content back to carbon dioxide. And while plants live, they are at the mercy of marauding animals which do not utilize atmospheric carbon dioxide but get their energy supplies by tearing down what the plants have built up. They *form* carbon dioxide as the result of their life processes and exhale it back into the atmosphere.

So there is a carbon dioxide cycle, with plants using it up and animals and bacteria forming it again. If animals gain a temporary ascendancy, plant life is killed off at too rapid a rate and enough animals starve to allow plants a chance to revive. If they revive too far, animals multiply in the lush environment and cut the plants down once more. So there are minor oscillations which (if never allowed to oscillate too far in either direction, and so far—knock wood—they haven't) average out, in the long run, into perfect balance.

Well, not in perfect balance. There are leaks in both directions.

For instance, some dead plant tissues don't get consumed by bacteria but get covered by muck and mire and clamped down underground where, under heat and pressure, the organic molecules are slowly stripped of everything but carbon and hydrogen, and sometimes all the way to carbon only. Thus oil and coal are formed and the carbon atoms contained therein are withdrawn permanently (or for hundreds of millions of years anyway) from the carbon dioxide pool of the air.

Also, carbon dioxide may react with the inorganic rocks to form insoluble carbonates and may be removed more or less permanently in that way.

Balancing both leaks out of the atmosphere is new carbon dioxide leaking into the atmosphere as the result of volcanic action.

With leaks in *both* directions, there is the possibility of balance still. At the present time, in fact, there is such a balance. About 15 to 30 million tons of carbon dioxide are removed permanently from the atmosphere each year as coal or insoluble carbonate. The same amount is restored each year by volcanic action. (Notice that the inorganic contribution to the cycle is not more than 0.05 per cent of the biochemical contribution. Here's an example of the importance of life on a planetary scale.)

Still, did the leaks always balance? After all, there may have been periods in Earth's history when the leak in one direction or another became particularly prominent. There were long periods of time when coal formation proceeded at an unusually high rate. The trillions of tons of coal that are buried underground have been withdrawn, however slowly, from the carbon dioxide pool of the air. Was that large-scale withdrawal replaced?

Again, during periods of mountain building, new rock is exposed to the atmosphere. Much more carbon dioxide than usual is used up in weathering and in the formation of insoluble carbonates. Is that carbon dioxide replaced?

On the other hand, there are periods of increased volcanic activity when more carbon dioxide is poured into the atmosphere than is true of most times.

Now then, does all this change the carbon dioxide content of the atmosphere from geologic era to geologic era? Probably yes, even if only slightly.

But if only slightly, does it matter? The answer to that is that some scientists think, yes, it matters a heck of a lot.

The major components of the atmosphere (oxygen and nitrogen) are, it seems, excellent transmitters of radiant energy over a broad stretch of wave lengths. The light rays of the Sun hit the air, pass through a hundred miles of it, hit the surface of the Earth, and are absorbed. The Earth heats up. The heated Earth radiates energy at night back into space, in the form of the far less energetic infra-red. This also passes through the atmosphere. The warmer Earth grows, the more heat is radiated away at night. At some particular equilibrium temperature, the net loss of radiation by Earth at night equals that gained by day so that, once the temperature (whatever it is) is reached, the Earth as a whole neither warms nor cools with time (barring internal radioactivity). Of course, individual portions of it may warm and cool with the seasons but this averages out, taken over the whole planetary surface.

Carbon dioxide, however, introduces a complication. It lets light rays through as easily as do oxygen and nitrogen, but it absorbs infra-red rather strongly. This means that

Earth's nighttime radiation finds the atmosphere partially opaque and some doesn't get through. The result is that the equilibrium temperature must rise a few degrees to reach the point where enough infra-red is forced out into space to balance the Solar input. The Earth is warmer (on the whole) than it would be if there were no carbon dioxide at all in the atmosphere. This warming effect of carbon dioxide is called the "greenhouse effect."

If there were a period of increased weathering or coal formation, so that the general carbon dioxide level of the atmosphere were to sink, the greenhouse effect would decrease and the earth's over-all temperature would drop. If volcanic action were to increase the carbon dioxide level, the over-all temperature would rise.

A recent set of calculations indicate that if the present carbon dioxide level should double, the over-all temperature of the Earth would rise by 3.6° C. If it were to halve, the temperature would drop 3.8° C.

Now to start an ice age going, you do not require a catastrophic temperature drop. The drop need just be enough to allow a little more snow to fall during the slightly colder winter than can be melted by the succeeding, slightly cooler summer. Repeat this year after year and the glaciers begin advancing. The chilled air and water drifting down from the North make the summers cooler than ever and the process accelerates.

The amount of temperature drop below the present level required to bring this about is not certainly known. Figures varying from a drop of 1.5 to 8° C. have been suggested. Adopting a middle-of-the-road position, cutting the carbon dioxide of the atmosphere in half (from 0.04 to 0.02 per cent) would drop the temperature 3.8° C. and that might well be enough to start an ice age. Perhaps such a change was actually the trigger of the ice ages that did happen.

A rise of 3 or 4°, on the other hand, would allow the slightly warmer summers to melt just a little bit more ice than can be replaced by the snows of the succeeding, slightly milder winters. The icecaps would melt and eventually disappear. There are currently 23 million cubic kilometers of ice in the world (mostly in Antarctica) and if this all melts, the volume of the oceans will increase by 1.7 per cent, the sea level would rise about 60 yards, and the coastal areas of the world would be flooded. (The Empire State Building would be in water to nearly the twentieth story.)

Obviously neither ice age nor world-wide tropic is desirable. Where we are is nice. Are we sure we are balanced or is there a slight trend one way or the other? Well, if there is, the trend ought to be so slight we need not worry for a million years—except for one thing.

Homo sapiens is throwing a monkey wrench into the machinery. We ourselves are upsetting the level by burning coal and oil in our, as aforesaid, nice, decent, non-radioactive, old-fashioned coal-burning (or oil-burning) power plants.

Until about 1900 the amount of carbon dioxide we formed in this manner was negligible. However, our industrialized twentieth century has been utilizing the "fossil fuels" in a logarithmically increasing fashion and the carbon dioxide that leaked out of the atmosphere over the space of a hundred million years of coal-forming is now being poured back into the atmosphere in a hundred million simultaneous puffs of smoke.

At the moment, we are adding 6 billion tons of carbon dioxide to the air each year (two hundred times as much as is being added by volcanic action and at least a fiftieth as much as is being added by life activity proper). And the rate is still increasing.

Even if we don't increase this rate, we will double the amount of carbon dioxide in the air (assuming there is no counteracting factor), raise the over-all temperature of the Earth 3.6° C., and make a healthy start at melting the icecaps *in toto* and drowning the coastal areas in a mere 350 years.

So much for our nice, decent, non-radioactive, old-fashioned coal-burning (or oil-burning) power plants.

Unless there is a counteracting factor. But is there?

Answer—maybe!

The first possibility is that as the atmospheric level of carbon dioxide goes up, plant life might luxuriate correspondingly, use the carbon dioxide faster, and bring the level down again. This would happen to begin with, probably. But then the natural interplay of life would balance that. More plants alive means more plants dying and decaying. It also means more animals to eat those plants. More decay and more animals means more carbon dioxide produced. The level would go back up again.

In other words, increasing the carbon dioxide of the air would speed up the carbon dioxide cycle but would not introduce a corrective influence. If we increased the carbon dioxide content of the air, it would stay increased, for all life processes could do.

But there is another factor. Leaving water vapor out of account, only one of the normal components of the air, carbon dioxide, is appreciably soluble in water. At $0°$ C., for instance, a milliliter (abbreviated ml.) of pure water will dissolve 0.0233 cubic centimeters (abbreviated cc) of nitrogen, and 0.0489 cc of oxygen; but it will dissolve 1.713 cc of carbon dioxide.

Now the oceans on Earth (which make up more than 98 per cent of Earth's total water supply) contain a total of 1.37×10^{24} ml. of an aqueous salt solution. If this all held carbon dioxide at the rate of 1.713 cc per ml. (so that the oceans fizzed like a planetful of soda water) there would be 2.35×10^{24} cc of carbon dioxide in solution. That would come, in weight, to 5.1×10^{15} tons, or about 2250 times as much carbon dioxide as there is in our entire atmosphere.

And actually this is a conservative estimate, since the solubility figures I gave are for pure water. This solubility goes up if the water is made alkaline, and sea water is indeed somewhat alkaline.

If the ocean can dissolve so much carbon dioxide, it seems odd that there remains any significant quantity of the gas in the atmosphere, unless the ocean happens to be saturated with it. It is nowhere near saturated, but the atmosphere retains the gas because the solution of carbon dioxide depends on a number of local factors (temperature, pressure, acidity, salinity, the life processes of ocean-dwelling organisms, etc.). Things are not as simple as though we put the oceans in a beaker and bubbled the atmosphere through it, stirring vigorously all the while.

By actual measurement, it has been estimated that the total carbon dioxide in the oceans is only 50 times that of the atmosphere.

Still, if this is the equilibrium, why shouldn't it be maintained when mankind goes about pouring carbon dioxide into the atmosphere while burning coal and oil. In other words, since 98 per cent of Earth's carbon dioxide is in the oceans, why shouldn't 98 per cent of Earth's *new* carbon dioxide go into the oceans.

If the ocean did in fact dissolve 98 per cent of new carbon dioxide as formed, the danger of tropicalization of Earth would recede. Instead of having the carbon dioxide level double, and Earth turn tropical, in 350 years, it would take

350 × 50 or 17,500 years to do so, and, heck, in that time, we'll think of something—we'll think of something—

However, the point of equilibrium is one thing and usually fairly easy to determine. The *rate at which equilibrium is reached* is quite another and often difficult to determine.

Yes, the ocean can dissolve the 6 billion tons of carbon dioxide we produce each year by burning coal and oil. There is plenty of room for it there. The ocean can hold 8 million times that quantity as a very minimum, over and above what it already holds. (This may create trouble for the fish, etc., but in eight million years we can solve that, perhaps.)

Nevertheless, though the oceans can dissolve it, will they do so quickly enough? If they will dissolve that quantity in a year, they keep pace with us, and all is well. If they dissolve it in a thousand years, we have produced 6000 billion tons of carbon dioxide (probably much more) meanwhile, and we are out of luck.

But then why shouldn't the ocean dissolve the carbon dioxide quickly? The gas is soluble enough and there's water enough and to spare in the oceans. What's to stop it?

Ah, you see the solution only takes place at the surface of the ocean where air and water meet. If the surface skin gets loaded with carbon dioxide, no more will dissolve. It won't matter that the water just under the skin is empty. The rate of solution will then depend on how fast the carbon dioxide molecules drift downward out of the skin, or how fast the ocean water moves about so that fresh water reaches the skin where it can dissolve additional carbon dioxide.

The latter process seems to solve the problem, since we all know that the ocean is always in a lashing turmoil. Surely, then, it is well-mixed, with new water reaching the surface all the time.

Right—if we consider only the top 600 feet of ocean. Just as all the storms of our atmosphere are confined to the troposphere (the lowest 5 to 10 miles) so all the wild water movements of the ocean are confined to the top 600 feet or less. Below 600 feet there is only a slow, majestic movement, exactly how slow and majestic we are not yet certain. The rate of carbon dioxide solution, then, depends on how quickly that deep water (representing about 94 per cent of the total ocean volume) is brought to the surface.

There is some sort of circulation between the depths and the surface, we know. After all, the ocean can't dissolve oxygen by any means more magical than those by which it dissolves carbon dioxide, and yet we know there is oxygen dissolved in the ocean all the way down to the lowest abysses. We know, because there is animal life in those abysses that could not live in the absence of oxygen.

The longer the water stays down in the depths without renewal, the lower the oxygen concentration becomes through consumption by living organisms. This offers one method of following water circulation in the abyss. Bring up samples of deep water from, say, 3 miles down, and measure the oxygen content. The higher the oxygen content, the more recently that water was at the surface.

Such measurements have been made, and it turns out that the deep water with highest oxygen content is in the North Atlantic and around Antarctica. Apparently that is where surface water sinks to the bottom most readily. On the bottom there seems to be a slow movement that carries the water out of the Atlantic, around Africa, into the Indian Ocean, through the South Seas and into the Pacific—with the oxygen content declining constantly.

Granted that such an abyssal circulation does exist, how fast does it move? We might find out by adding something to the top of the ocean which is not already present in the ocean. Then we would have to wait for it to show up in various portions of the abyss and note the elapsed time. Of course, the added something would have to be detectable in extremely tiny quantities after we have allowed for dilution by an entire oceanful of water.

Actually, there may be something that fits the bill—strontium-90. There is a detectable quantity in the atmosphere now and there wasn't any fifteen years ago. Some has gotten into the ocean's skin but is there any already in the deep waters? If so, where? Chemists are devising methods of concentrating and measuring the strontium-90 in the ocean for just this purpose.

It would be odd if it turned out that the dangerous ash, strontium-90, were to give us vital information involving the dangers of the "harmless" ash, carbon dioxide. It's an ill wind——

(The abyssal circulation is important not only with respect to information concerning the carbon dioxide cycle. The lower waters are richer in minerals—hence more fertile—than the life-scavenged upper waters. If the time comes when man depends on the sea for most of his food, a knowledge of abyssal circulation may be vital for "ocean-farming.")

Of course, why theorize as to how the ocean *may* dissolve carbon dioxide, how slowly the atmospheric carbon dioxide *may* be building up, how quickly the Earth *may* be turning into an iceless tropical world. Why not actually measure the icecaps of the world and see if they are disappearing or not. And if they are disappearing, how quickly? This, in fact, was one of the prime objects of research for the Geophysical Year and one of the most important reasons for all those scientists setting up housekeeping on the Antarctic icecap.

We might also measure the actual over-all temperature of the Earth and see if it is going up. If all the combusted carbon dioxide stays in the atmosphere, while dissolving in the oceans at only a negligible rate, then the over-all temperature ought to go up $1.1°$ C. per century.

According to Gilbert N. Plass of Johns Hopkins, such temperature measurements as are available indicate that just this rate of temperature increase has indeed been going on since 1900. Of course, temperature measurements during the first half of the twentieth century are not reliable outside the more industrialized countries, so maybe this apparent increase matches the theoretical only through a coincidence arising from insufficient data.

However, if this is more than coincidence; if Earth is really warming up at that rate, then wave good-by to the icecaps. And if you live at the seashore, your not too distant descendants may well have to visit the old homestead with a skin-diver's outfit.

Earth has survived a similar fate three times in the last 300,000 years, this current rise being the fourth. These periods of rise make up what are called the "interglacial epochs." Earth has also survived four periods of temperature drop in this same period of time, each of which initiated a "glacial epoch" or, as it is more commonly known, an "ice age." It might seem that there is some physical phenomenon which brings on this coming and going of the ice, and one would expect that same phenomenon to continue and to keep the oscillation of ice and no-ice going for the immediate future (by which I mean the next few million years).

Yet prior to 300,000 years ago (for at least 200,000,000 years prior, in fact) there were no Ice Ages. For all that long period (or more) Earth was reasonably ice-free. Naturally, the question arises: what happened 300,000 years ago?

One explanation is that Earth undergoes a temperature oscillation of a very slow and majestic type which didn't make itself visible in the form of ice till 300,000 years ago. For instance, a Serbian physicist named Milutin Milankovich in the 1920s suggested that because of oscillations in Earth's orbit and the tilt of its axis, the planet picks up a bit more heat from the Sun at some times than at others. His proposed temperature cycle lasted 40,000 years, so that there is a kind of 20,000-year long "Great Summer" and a 20,000-year long "Great Winter." The temperature differences involved are not really very great but, as I stated earlier a drop of less than 4° C. in Earth's present temperature would be enough to kick off an Ice Age.

This Milankovich-oscillation can be made to explain the recent advances and retreats of the glaciers, but what about the situation B.I.A. (Before the Ice Ages)?

Well, what if prior to 300,000 years ago, Earth's over-all temperature were sufficiently high so that even the Great Winter dip was not enough to bring on the ice? You can see that, if you consider the annual temperature oscillation between ordinary summer and winter. In New York this oscillation crosses the freezing point of water, so there is rain in the summer but snow in the winter. In Miami the average temperature is higher and the oscillation does not dip low enough to bring snow in the winter. On a planetary scale, what if Earth's climate switched from iceless Miami to periodically icy New York?

This possibility has been checked by isotope analysis. (These days, if a scientist can't get an answer by isotope analysis, he ain't hep.) There are three stable oxygen isotopes: oxygen-16, which makes up 99.76 per cent of all the oxygen atoms; oxygen-18 (0.20 per cent) and oxygen-17 (0.04 per cent). They all behave almost alike, so alike that ordinarily no difference can be detected. However, oxygen-18 is 12½ per cent heavier than oxygen-16 and correspondingly slower in its reactions. For instance, when water evaporates, water molecules containing oxygen-16 get into the air a bit more easily than those containing oxygen-18, and if evaporation continues over a long interval the water that is left contains noticeably more oxygen-18 than it had originally.

This applies to the oceans, which are constantly evaporating, so that sea water should (and does) have a bit more oxygen-18, in proportion to oxygen-16, than does fresh water, which is made up of the evaporated portion of the oceans. Furthermore, this effect is increased as the temperature goes up. For each 1° C. rise in the temperature of the ocean, the ratio of oxygen-18 to oxygen-16 goes up 0.02 per cent.

Now then, fossil sea shells are made up largely of calcium carbonate. The calcium carbonate contains oxygen atoms which were derived from the ocean water. The oxygen-18/oxygen-16 ratio in those shells must therefore reflect the ratio in the water from which they derived the oxygen and that, in turn, should give a measure of the ocean temperatures of ages long past.

Such measurements were first made in the laboratories of Harold C. Urey at the University of Chicago and proved a very tricky job. On the basis of such measurements, however, it turns out that during the Mesozoic Age of old, when dinosaurs were bold, the ocean temperatures were as high as 21° C. (70° F.).

This bespeaks a planetary temperature too high to allow an Ice Age, even at the bottom of the Milankovich cycle.

But then, beginning 80,000,000 years ago, when ocean temperatures were at the 21° peak, the temperatures started dropping and have continued to do so ever since.

According to Cesare Emiliani (who carried temperature measurements into the recent past, geologically speaking) the explanation for this is that, after a long period of land area fairly free of mountains and oceans fairly free of abyss, so that many shallow seas covered much of what is now land, a geological revolution occurred. The ocean bottoms started sinking and mountain ranges started rising.

With land going up and ocean bottoms down, new land was exposed very gradually. Land stores less heat than does water, radiating more away at night, so that the Earth's over-all temperature gradually dropped. Also, new land meant new rocks exposed to carbon dioxide weathering, which meant a fall in the carbon dioxide of the atmosphere, a decrease of the "greenhouse effect," and again, a fall in temperature.

Quite possibly it was this fall in temperature that killed off the dinosaurs.

By a million years ago, the steady drop of ocean temperatures had brought it down to 2° C. (35½° F.) and by 300,000 years ago, Earth's temperature was low enough to allow the Ice Ages to appear at the bottom of the Milankovich cycles.

A somewhat more startling explanation of the beginning of the Ice Ages has been advanced by Maurice Ewing and

William Donn, working at Columbia. They blame it specifically on the Arctic Ocean.

The North Pole is located in a small, nearly landlocked arm of the ocean, which is small enough and landlocked enough to make possible an unusual state of affairs.

Thus the suggestion is that when the Arctic Ocean is free of ice, it acts as a reservoir of evaporating water that feeds snowstorms in the winter. If the Arctic Ocean were large and open, most of these snow storms would fall upon the open sea and there melt. As it is, the snow falls upon the surrounding land areas of Canada and Siberia, and because of the lower heat content of land areas, it does not melt but remains during the winter. In fact, it accumulates from winter to winter, with the summer never quite melting all the ice produced by the preceding winter. The glaciers form and creep southward.

Once this happens, a considerable fraction of the Earth is covered with ice, which reflects more of the Sun's radiation than does either land or water. Furthermore, the Earth as a whole is cloudier and stormier during an Ice Age than otherwise, and the excess clouds also reflect more of the Sun's radiation. Altogether, about 7 per cent of the Sun's radiation, that would ordinarily reach Earth, is reflected during an Ice Age. The Earth's temperature drops and the Arctic Ocean, which (according to Ewing and Donn) remained open during the height of glacier activity, finally freezes over. (Even despite the lowering temperature, it does this only because it is so small and landlocked.)

Once the Arctic freezes over, the amount of evaporation from it is drastically decreased, the snowstorms over Canada and Siberia are cut down, the summers (cool as they are) suffice to melt more than the decreased accumulation, and the glaciers start retreating. The Earth warms up again (as it is now doing), the Arctic Ocean melts (this point not yet having been reached in the current turn of the cycle), the snows begin again, and bang comes another glaciation.

But why did all this only start 300,000 years ago? Ewing and Donn say because that is when the North Pole first found itself in the Arctic Ocean. Before then it had been somewhere in the Pacific where the ocean was large enough and open enough to cause no severe snowstorms on the distant land areas.

Ice Ages could continue to annoy us periodically, then, until the present mountains wear down to nubs and the ocean bottoms rise, or until the North Pole leaves the Arctic (depending on which theory—if either—is correct).

Unless, that is, something new interferes, such as the carbon dioxide we are pouring into the atmosphere. The current temperature rise is being radically hastened, apparently, by the increased carbon dioxide in the atmosphere. The next temperature drop may be correspondingly slowed and may, conceivably, not drop far enough to start a new glaciation.

Therefore, it is possible that Earth has seen its last Ice Age, regardless of the Milankovich cycle or the position of the North Pole, until such time as the ocean, or we ourselves, can get rid of the excess carbon dioxide once again. Within a matter of centuries, then, we may reverse much or all of an 80,000,000 year trend of dropping temperature and find ourselves back in the Mesozoic, climatically speaking, only without (thank goodness) the dinosaurs.

3 Thin Air

Earth's atmosphere is now going through a period of scientific importance and prominence. To put it as colorfully (and yet as honestly) as possible, it is all the scientific rage.

Once before in scientific history, Earth's atmosphere passed through a period of glamour. Let me tell you about that before I get to the current period.

To begin with, in ancient Greek times, air had all the dignity of an "element"; one of the abstract substances out of which the Universe was composed. According to the philosophers, culminating in Aristotle, the Universe was composed of "earth," "water," "air," and "fire" in four concentric shells, with earth at the bottom and fire at the top.

In modern terms, earth is equivalent to the lithosphere, the solid body of the planet itself. Water is the hydrosphere, or ocean; and air is the atmosphere. Fire is less obvious, being so high (according to Aristotle) as to be ordinarily imperceptible to human senses. However, storms roiled the sphere of fire and made fragments of it visible to us as lightning.

Even the sphere of fire reached only to the Moon. From the Moon outward, there was a fifth and heavenly "element," like none of those on our imperfect earth. Aristotle called it "ether." Medieval scholars called it "fifth element" but did so in Latin, so that the word came out "quintessence." That word survives today, meaning the purest and most essential part of anything.

Such a theory as to the structure of the universe presented early thinkers with few problems about the air. For instance, did the atmosphere ever come to an end as one went upward? Sure it did. It came to an end at the point where the sphere of fire began.

34

You see, there was always *something* in the Aristotelian view. Just as earth gave way to water and water to air, with no gap between, so air gave way to fire and fire to ether. There was never *nothing*. As the ancient scholars said, "Nature abhors a vacuum."

Did the atmosphere weigh anything? Obviously not. You didn't feel any weight, did you? If a rock fell on you or a bucket's worth of water, you would feel the weight. But there's no feeling of weight to the air. Aristotle had an explanation for this. Earth and water had a natural tendency to move downward, as far as they could, toward the center of the universe (i.e. the center of the Earth).

Air, on the other hand, had a natural tendency to move upward, as anyone could plainly see. (Blow bubbles under water and *watch* them move upward—not that Aristotle would appeal to experiment, believing as he did that the light of reason was sufficient to penetrate the secrets of nature.) Since air lifted upward, it had no weight downward.

Aristotle flourished about 330 B.C. and his views were gospel for a long time.

Curtain falls. Two thousand years pass. Curtain rises.

Toward the end of his long and brilliant life, Galileo Galilei, the Italian scientist, grew interested in the fact that an ordinary water pump drawing water out of a well would not lift the water any higher than about 33 feet above the natural level. This no matter how vigorously and how pertinaciously the handle of the pump was operated.

Now people thought they knew how a pump worked. It was so designed that a tightly fitted piston moved upward within a cylinder, creating a vacuum. Since Nature abhorred a vacuum, water rushed upward to fill said vacuum and was trapped by a one-way valve. The process was repeated and repeated, more and more water rushed upward until it poured out the spout. Theoretically, this should go on forever, the water rising higher and higher as long as you worked the pump.

Then why didn't water rise more than 33 feet above its natural level? Galileo shook his head, and never did find an answer. He muttered gruffly that apparently Nature abhorred a vacuum only up to 33 feet and recommended that his pupil Evangelista Torricelli look into the matter.

In 1643, the year after Galileo's death, Torricelli did that. It occurred to him that what lifted the water wasn't a fit of emotion on the part of Dame Nature, but the very unemotional weight of air pressing down on the water and forcing it upward into a vacuum (which would ordinarily be filled with a balancing weight of air). Water could not be forced higher than 33 feet because a column of water 33 feet high pressed down as hard as did the entire atmosphere, so that there was a balance. Even if a complete vacuum were pulled over the water, so that air down at well-water level pushed the column upward without any back air pressure, the weight of the water itself was enough to balance the total air pressure.

How to test this? If you could start with a column of water, say, 40 feet long, it should sink until the 33-foot level was reached. A 40-foot column of water would have more pressure at the bottom than the entire atmosphere. But how to handle 40 feet of water?

Well then, suppose you used a liquid denser than water. In that case, a shorter column would suffice to balance air's pressure. The densest liquid Torricelli knew of was mercury. This is about 13½ times as dense as water. Since 33 divided by 13½ is about 2½ feet, a column of 30 inches of mercury should balance the air pressure.

Torricelli filled a tube (closed at one end and a yard long) with mercury, put his thumb over the open end, and tipped it into an open container of mercury. If the air had no weight, it would not press on the exposed mercury level in the container. All the mercury in the tube would therefore pour out.

The mercury in the tube started pouring out, to be sure, but only to the extent of a few inches. Fully 30 inches of mercury remained standing, supported by nothing, apparently. It was either magic or else Aristotle was wrong and air had weight. There was no choice; air had weight. Thus the first glamorous period of the atmosphere had begun.

Torricelli had invented the barometer, an instrument still used today to measure air pressure as so many inches of mercury. Furthermore, in the upper part of the tube, in the few inches that had been vacated by the mercury, there was a vacuum, filled with nothing but some mercury vapor and darned little of that. It is called a Torricellian vacuum to this day and was the first decent man-made vacuum ever formed. It showed definitely that Nature didn't give a plugged nickel one way or the other for vacuums.

In 1650 Otto von Guericke, who happened to be mayor of the German city of Magdeburg, went a step further. He invented an air pump which could pump air out of an enclosure, forming a harder and harder vacuum; i.e. one that grew more and more vacuous.

Von Guericke then demonstrated the power of air pressure in a dramatic way. He had two metal hemispheres made which ended in flat rims that could be greased and stuck together. If this were done, the heavy hemispheres fell apart of themselves. There was nothing to hold them together.

But one of the hemispheres had a valved nozzle to which an air pump could be affixed. Von Guericke put the hemispheres together and pumped the air out of them, then closed the valve. Now the weight of the atmosphere was pressing the hemispheres together and there was no equivalent pressure within.

How strong was this air pressure? Well, publicity-wise von Guericke attached a team of horses to one hemisphere by a handle he had thoughtfully provided upon it and another team to the other hemisphere. With half the town of Magdeburg watching open-mouthed, he had the horses strain uselessly in opposite directions.

The thin air about us which "obviously" weighed nothing did indeed weigh plenty. And when that weight was put to use, two teams of horses couldn't counter it.

Von Guericke released the horses, opened the valve, and the hemispheres fell open by themselves. It was as dramatic an experiment as Galileo's supposed tossing of two balls of different mass off the Tower of Pisa, and what's more, von Guericke's experiment really happened. (They don't make mayors like that anymore.)

Since the atmosphere has weight, there could only be so much of it and no more. There could only be enough of it to allow a column of air (from sea level to the very tiptop), with a cross-sectional area of one square inch, to weigh 14.7 pounds. If the atmosphere were as dense all the way up as it is at sea level, a column just five miles high would have the necessary weight.

But of course, air isn't equally dense all the way up.

In the 1650s a British scientist, Robert Boyle, having read of von Guericke's experiments, set about to study the properties of air more thoroughly. He found it to be compressible.

That is, if he trapped a sample of air in the short closed half of a U-tube by pouring mercury into the long, open half, the trapped air contracted in volume (i.e. was compressed) until it had built up an internal pressure that balanced the head of mercury. As the mercury was added or removed the trapped air compressed and expanded like a spring. The English scientist, Robert Hooke, had just been reporting on the behavior of actual springs and since the trapped air behaved analogously, Boyle called it "the spring of the air."

If, now, Boyle poured additional mercury into the U-tube, the trapped air decreased further in volume until the internal pressure had increased to the point where the additional weight of mercury could be supported. Furthermore, Boyle made actual measurements and found that if the pressure on the trapped air was doubled, its volume was halved; if the pressure was tripled, the volume was reduced to one third and so on. (This is one way of stating what is now called Boyle's law.)

This was a remarkable discovery, for liquids and solids did not behave in this way. Boyle's work marks the beginning of the scientific study of the properties of gases which, in a hundred years, produced the atomic theory and revolutionized chemistry. This was just another consequence of this first glamorous period of the atmosphere.

Since air is compressible, the lowest regions of the atmosphere, which bear all the weight of all the air above must be most compressed. As one moves upward in the atmosphere, each successive sample of air at greater and greater heights has less atmosphere above it, is subjected to a smaller weight of air, and is therefore less compressed.

It follows that a given number of molecules takes up more room ten miles up than they do at sea level, and more room still twenty miles up and more room still thirty miles up and so on indefinitely. From this, it would seem that the atmosphere must also stretch up indefinitely. True, there's less and less of it as you go up, but that less and less is taking up more and more room.

In fact, it can be calculated that, if the atmosphere were at the sea-level average of temperature throughout its height, air pressure would be reduced tenfold for every twelve miles we travel upward. In other words, since air pressure is 30 inches of mercury at sea level, it would be 3 inches of mercury at a height of 12 miles, 0.3 inches of mercury at 24 miles, 0.03 inches of mercury at 36 miles, and so on.

Even at a height of 108 miles, there would still be, by this accounting, 0.00000003 inches of mercury of pressure. This doesn't sound like much, but it means that six million tons of air would be included in the portion of the atmosphere higher than 100 miles above Earth's surface.

Of course, the atmosphere is *not* the same temperature

throughout. It is the common experience of mankind that mountain slopes are always cooler than the valley below. There is also no denying the fact that high mountains are perpetually snow-covered at the top, even through the summer and even in the tropics.

Presumably, then, the temperature of the atmosphere lowers with height and, it seemed likely, did so in a smooth fall all the way up. This spoiled the simple theory of decline of density with height but it didn't alter the fact that the atmosphere was remarkably high. Once astronomers started looking, they found ample evidence of that.

For instance, visible meteor trails have been placed (by triangulation) as high as 100 miles. That means that even at 100 miles, then, there is enough atmosphere to friction tiny bits of metals to incandescence.

Furthermore, aurora borealis (caused by the glowing of thin wisps of gas as the result of the bombardment with particles from outer space) has been detected as high as 600 miles.

However, how was one to get details on the upper atmosphere? Particularly one would want to know the exact way in which temperature and pressure fell off with height. As early as 1648 the French scientist, Blaise Pascal, had sent a friend up a mountain side with a barometer to check the fall of air pressure; but then, how high are mountains?

The highest mountains easily accessible to the Europeans of the seventeenth century were the Alps, the tallest peaks of which extend 3 miles into the air. Even the highest mountains of all, the Himalayas, only double that. And then, how could you be sure that the air 6 miles high in the Himalayas was the same as the air 6 miles high over the blank and level ocean.

No, anything in the atmosphere higher than, say, a mile was attainable only in restricted portions of the globe and then with great difficulty. And anything higher than 5 or 6 miles just wasn't attainable, period. No one would ever know. No one.

So the first glamorous period of the atmosphere came to an end.

Curtain falls. A century and a half passes. Curtain rises.

In 1782 two French brothers, Joseph Michel Montgolfier and Jacques Étienne Montgolfier, lit a fire under a large light bag with an opening underneath and allowed the heated air and smoke to fill it. The hot air, being lighter than the cold air, moved upward, just as an air bubble would move upward in water. The movement carried the bag with it, and the first balloon had been constructed.

Within a matter of months, hydrogen replaced hot air, gondolas were added, and first animals and then men went aloft. In the next few decades, aeronautics was an established craze—a full century before the Wright brothers.

Within a year of the first balloon, an American named John Jeffries went up in one, carrying a barometer and other instruments, plus provisions to collect air at various heights. The atmosphere, miles high, was thus suddenly and spectacularly made available to science and the second glamorous period had begun.

By 1804 the French scientist, Joseph Louis Gay-Lussac, had gone up nearly 4½ miles in a balloon, a height considerably greater than that of the highest peak of the Alps, and brought down air collected there.

It was, however, difficult to go much higher than that, because the aeronauts had the inconvenient habit of breathing. In 1874, three men went up 6 miles—the height of Mount Everest—but only one survived. In 1892 the practice of sending up unmanned (but instrumented) balloons was inaugurated.

The most important purpose of the early experiments was the measurement of the temperature at heights and by the 1890s some startling results showed up. The temperature did indeed drop steadily as one went upward, until at a height somewhat greater than that of Mount Everest, the temperature of –70° F. was reached. Then, for some miles higher, *there were no further temperature changes.*

The French meteorologist, Leon P. Teisserenc de Bort, one of the discoverers of this fact, therefore divided the atmosphere into two layers. The lower layer, where there was temperature change, was characterized by rising and falling air currents that kept that region of the atmosphere churned up and produced clouds and all the changing weather phenomena with which we are familiar. This is the *troposphere* ("the sphere of change").

The height at which the temperature fall ceased was the *tropopause* ("end of change") and above it was the region of constant temperature, a place of no currents or churning, where the air lay quietly and (Teisserenc de Bort thought) in layers, with the lighter gases floating on top. Perhaps the earth's atmospheric supply of helium and hydrogen were to be found up there, floating on the denser gases below. He called this upper layer the *stratosphere* ("sphere of layers").

The tropopause is about ten miles above sea level at the equator and only five miles above at the poles. The stratosphere extends from the tropopause up to about sixteen miles. There, where the temperature starts changing again, is the *stratopause*.

About 75 per cent of the total air mass of the earth exists within the troposphere and another 23 per cent is in the stratosphere. Together, troposphere and stratosphere, with 98 per cent of the total air mass between them, make up the "lower atmosphere." But it is the 2 per cent above the stratopause, the "upper atmosphere," which gained particular prominence as the twentieth century wore on.

In the 1930s ballooning entered a new era. Balloons of polyethylene plastic were lighter, stronger, less permeable to gas than the old silken balloons (cheaper, too). They could reach heights of more than twenty miles. Sealed gondolas were used and the balloonists carried their own air supply with them.

In this way, manned balloons reached the stratosphere and beyond. Russian balloonists brought back samples of stratospheric air and no helium or hydrogen was present; just the usual oxygen and nitrogen. (We now know that the atmosphere is largely oxygen and nitrogen all the way up.)

Airplanes with sealed cabins were flying the stratosphere, too, and toward the end of World War II, the *jet streams* were discovered. These were two strong air currents girdling the earth, and moving from east to west at 100 to 500 miles per hour at about tropopause heights, one in the North Temperate Zone and one in the South Temperate. Apparently they are of particular importance in weather forecasting, for they wriggle about quite a bit and the weather pattern follows their wriggling.

After World War II, rockets began going up and sending down data. The region above the stratosphere was more and more thoroughly explored. Thus it was found that from the stratopause to a height of about 35 miles, the temperature *rises*, reaching a high of –55° F. before dropping once more to –100° F. at a height of about 50 miles. Above that there is a large and steady rise to temperatures that are estimated to be about 2200° F. at a height of 300 miles and are probably higher still at greater heights.

The region of rising, then falling, temperature, from 16 to

50 miles is now called the *mesosphere* ("the middle sphere") and the region of minimum temperature that tops it is the *mesopause*. The mesosphere contains virtually all the mass of the upper atmosphere, about 2 per cent of the total. Above the mesopause, only a few thousandths of a per cent of the atmosphere remain.

These last wisps are, however, anything but insignificant, and they are divided into two regions. From 50 to 100 miles is the region where meteor trails are visible. This is the *thermosphere* ("sphere of heat" because of the rising temperatures) and is topped by the *thermopause* though that is *not* the "end of heat." Some authorities run the thermosphere up to 200 or even 300 miles.

Above the thermopause, is the region of the atmosphere which is too thin to heat meteors to incandescence but which can still support the aurora borealis. This is the *exosphere* ("outside sphere").

There is no clear upper boundary of the exosphere. Actually, the exosphere just thins and fades into interplanetary space (which is *not*, of course, a complete vacuum). Some try to judge the "end of the atmosphere" by the manner in which the molecules of the air hit one another.

Here at sea level, molecules are crowded so closely together that any one molecule will only be able to travel a few millionths of an inch (on the average) before striking another. The air acts like a continuous medium, for that reason.

At a height of ten miles, the molecules have so thinned out that they may travel a ten-thousandth of an inch before colliding. At a height of 70 miles, they will travel a yard and a half and at 150 miles, 370 yards before colliding. At a height of several hundred miles, collisions become so rare that you can ignore them and the atmosphere begins to behave like a collection of independent particles.

(If you have ever been part of the New Year's Eve crowd in Times Square, and have also walked a lonely city street at 2 A.M., you have an intuitive notion of the difference between particles composing an apparently continuous medium and particles in isolation.)

The point where the atmosphere stops behaving like a continuous medium and begins to act like a collection of independent particles may be considered the *exopause*, the end of the atmosphere. This has been placed at heights varying from 600 to 1000 miles by different authorities.

The practical importance to us of the upper atmosphere is that it bears the brunt of the various bombardments from outer space, blunting them and shielding us.

For one thing there is the Sun's heat. The Sun emits photons with the energy one would expect of a body with the surface temperature of 10,000° F. These photons do not lose energy as they travel through space, and consequently strike the atmosphere in full force. Fortunately, the Sun radiates them in all directions and only a billionth or so are intercepted by our own planet.

Still, when one of the photons strikes a molecule at the edge of the atmosphere and is absorbed, that molecule may find itself possessed of a Sun-type temperature of 10,000° F. Only a small proportion of the molecules of Earth's atmosphere are so heated and slowly, by collision with other molecules below, the energy is shared so that the temperature drops to bearable levels as one descends.

(The high temperatures of the exosphere and thermosphere are an odd echo of the Aristotelian sphere of "fire." You may also be wondering how rockets can pass through the exosphere, if it has a temperature in the thousands of degrees, without being destroyed. There you run up against the difference between temperature and heat. The individual molecules have much energy, i.e. have a high temperature, but there are so few of them, that the total energy, i.e. heat, is negligible.)

Of course, the high temperature of the outermost atmosphere has its effects on the molecules that compose it. Oxygen and nitrogen molecules, shaken by this temperature and exposed to the bombardment of high energy particles beside, break up into individual atoms. (If the free atoms sink down to positions where less energy is available, they recombine, so no permanent damage is done.)

People have wondered whether ram-jets might not make use of these free atoms to navigate the exosphere. If enough could be gathered and compressed (and that is the hard part) the energy delivered per weight by their reunion to form molecules would be much higher than the energy delivered per weight by the combination of conventional fuels with oxygen, ozone, or fluorine.

Furthermore, the supply would be inexhaustible, since the atoms, once combined into molecules, would be expelled out the rear where the Sun's energy would promptly split them into atoms again. In effect, such a ram-jet would be running on solar energy, one tiny step removed.

The bombardment of particles from space also succeeds in damaging individual atoms or molecules, knocking off one or more planetary electrons, and leaving behind charged atom fragments called *ions*. Enough ions are formed in the exosphere to produce the glow called the auroras.

In the denser air of the thermosphere, there are more or less permanent layers of ions at different heights. These first made themselves known by the fact that they reflect certain radio waves. In 1902 Oliver Heaviside of England and Arthur Edwin Kennelly of the United States discovered (independently) the lowest of these layers, about 70 miles high. It is called the Kennelly-Heaviside layer in their honor.

Higher layers (at about 120 miles and 200 miles) were discovered in 1927 by the British physicist, Edward Victor Appleton, and these are called the Appleton layers. Because of these various layers of ions, the thermosphere is frequently called the *ionosphere,* and its upper boundary the *ionopause* (though that is not the "end of ions" any more than the "end of heat").

Nowadays, the layers have received objective letters. The Kennelly-Heaviside layer is the E layer, while the Appleton layers are the F_1 layer and F_2 layer. Between the F_1 layer and the E layer is the E region and below the E layer is the D region.

Yet lower in the atmosphere, down in the mesosphere, the ultraviolet of the Sun is still capable of inducing chemical reactions that do not ordinarily proceed spontaneously at sea level. It is possible to send chemicals up there and watch things happen. In the main, though, the important point is that something happens to a chemical already present there. Ordinary oxygen molecules of the mesosphere (made up of two oxygen atoms apiece) are converted into the more energetic ozone molecules (made up of three oxygen atoms apiece).

The ozone is continually changing back to oxygen while the forever incoming ultraviolet is continually forming more ozone. An equilibrium is reached and a permanent layer of ozone exists about 15 miles above the Earth's surface. This is fortunate for us since the maintenance of the ozone layer continually absorbs the Sun's hard ultraviolet which, if it were allowed to reach the Earth's surface unabsorbed, would be fatal for most forms of life in short order.

Because of the chemical reactions proceeding in the mesosphere it is sometimes called the *chemosphere* (and its upper boundary, the *chemopause*). As for the ozone layer itself, that is sometimes referred to as the *ozonosphere*.

So there you have the steps. From Aristotle's undifferentiated "air" through one period of scientific glamour to Boyle's smoothly thinning atmosphere; then through another period of scientific glamour to the modern layers upon layers of air, with changing properties.

Next step (now begun): the investigation of *cis-Lunar space* (the space "this side of the Moon") which has already yielded the surprising knowledge of the existence of the Van Allen radiation belts—and what else?

Well, wait and see.

4 Catching Up with Newton

It is very irritating that, in this modern era of missiles and satellites, there are so many newsmen who haven't caught up with Newton yet. They speak with appalling glibness about the weightlessness experienced by a spaceman once he has climbed "beyond the reach of gravity." Presumably they have the impression that there is a boundary line near the top of the atmosphere or thereabouts, beyond which there is suddenly no gravity—and that is the very thing Newton's theory disallows.

Isaac Newton was the first to formulate the Law of Universal Gravitation. Note the adjective "Universal," which is the important word. Newton did *not* discover that apples fell to the ground when they broke loose from the tree; that was common knowledge. What he did demonstrate was that the Moon's path around the Earth could be explained by supposing that the Moon was in the grip of the same force that tugged at the apple.

His great suggestion was that every piece of matter in the Universe attracted every other piece of matter, and that the quantity of this force could be expressed in a simple formula.

The force of attraction (f) between any two bodies, said Newton, is proportional to the product of the masses (m_1 and m_2) of the bodies and inversely proportional to the square of the distance (d) between their centers. By introducing a proportionality constant (G), we can set up an equation representing the above statement symbolically:

$$f = Gm_1m_2/d^2 \qquad \text{(Equation 1)}$$

The most recent and presumably most accurate value obtained for G (in 1928, at the Bureau of Standards) is 6.670

$\times 10^{-8}$ dyne cm^2/sec^2. This means that if two 1-gram spherical masses are placed exactly 1 centimeter apart (center to center), the attraction between them is 6.670×10^{-8} dynes.

This shows gravity to be a relatively weak force as compared with electrical and magnetic attractions, for instance. One dyne of force is equivalent, roughly, to 1 milligram of weight. If the two 1-gram spheres were the only matter in the universe, therefore, each would weigh, under the gravitational attraction of the other at the distance indicated, only 0.000000066 milligrams (or about two trillionths of an ounce). However, when masses as large as the Earth are concerned, even a weak force like gravity becomes tremendous.

Of course, we don't have to use dynes or any other fancy units to understand the essentials of gravity. Suppose, for instance, that the two masses between which we are trying to measure gravitational attraction are a spaceship and the planet Earth. The mass of the spaceship we can set equal to 1 (one what? one spaceship-mass). The mass of the Earth we can also set equal to 1, by using different units—one Earth-mass, this time.

The distance between the center of the Earth and the center of the spaceship, which we will suppose to be resting on the Earth's surface, is just about 3950 miles. We can make this value also 1 by calling that number of miles 1 Earth-radius.

Notice, now, that in using Newton's equation, it is necessary to take distances from center to center. In other words, the important point is not how far the spaceship is from the surface of the Earth, but how far from its center.

It is one of Newton's great accomplishments, you see, that he was able to demonstrate that spheres of uniform density attract each other as though all their mass were concentrated at the central point. To be sure, actual heavenly bodies are not uniformly dense, but Newton also showed this central-point business to be true for spheres which consisted of a series of layers (like an onion) each of which was uniform in density, though the density might vary from layer to layer. This modified situation *does* hold true for actual heavenly bodies.

But back to Earth and spaceship. Now that we have chosen convenient units for masses and distances, it is only necessary to make the gravitational constant also 1 (one constant-value) and Equation 1 becomes:

$$f = 1 \times 1 \times 1/1^2 \qquad \text{(Equation 2)}$$

Therefore, as the result of our shrewd unit choices, it turns out that the force of attraction between Earth and space-ship is exactly 1.

So far so good, but this is for the spaceship resting on Earth's surface. What if it were not on Earth's surface but 3950 miles straight up?

By changing the spaceship's position, we are not altering its mass, or Earth's mass or the gravitational constant. Each of these can remain 1. The only thing that is being altered is the distance between the center of the spaceship and the center of the Earth, so distance is all we need concern ourselves with and Equation 2 becomes:

$$f = 1/d^2 \qquad \text{(Equation 3)}$$

Now, then, if the spaceship is 3950 miles above the Earth's surface, its distance from the center of the Earth is 3950 miles plus 3950 miles or 2 Earth-radii. (We can use any units we want but, once having chosen them, we must stick with them. Such are the ethics of the situation.)

At 3950 miles above the Earth's surface, then, the force of attraction between Earth and the spaceship, using Equation 3, is $1/2^2$ or 0.25.

Gravitational attraction is usually measured by weighing an object. Consequently we can say that whatever the weight of the spaceship on the surface of the Earth, it weighs (i.e. is attracted by the Earth) only ¼ as much 3950 miles above the surface.

By the same reasoning we could show that this would hold for any object other than the spaceship. The gravitational attraction of the Earth for anything at all drops to a quarter of its value as that "anything at all" is moved from Earth's surface to a height 3950 miles above its surface.

Equation 3, will also give us the force between the Earth and the spaceship (or any other object) for any height above the surface. Some figures, so obtained, are shown in Table 1.

As you see, gravitational force starts dropping off at once. Even at low satellite-heights, so to speak, it varies from ⅔ to ⁹⁄₁₀ of what it is at the planet's surface. Or, to get really petty about it: if you weigh 150 pounds and are suddenly transported to the top of Mount Everest from your sea-level home, you would find gravity weakened enough to make your weight 149½ pounds.

Nevertheless, Earth's gravitational force does not drop to zero, no matter what the distance. No matter how large you made d in Equation 3, f is never zero. If you go back to Equation 1, you would see that this is also true for the attraction between any two bodies, however small, with masses greater than zero. In other words, the gravitational influence of every body, however small, is exerted through all of space.

Nor does the force very quickly become negligible when large bodies are involved. The gravitational force between Earth and Venus, at closest approach is only 0.000000025 that of what it would be if the two planets were in contact. Nevertheless, the force attracting the Earth and Venus, even at a distance of 25,000,000 miles is still equal to 130 trillion tons.

So much for spacemen getting "beyond the reach of gravity."

The word "Universal" in Newton's law wouldn't be worth much, if we don't apply the equation to other bodies. We can start by supposing the spaceship to be resting on the surface of the Moon.

To begin with, the spaceship has the same mass (i.e. the quantity of matter contained in its substance) as on Earth and we agreed to let that mass (m_1) equal 1. The constant G never varies and we agreed to let that equal 1, also. Equation 1 therefore becomes:

$$f = m_2/d^2 \qquad \text{(Equation 4)}$$

where m_2 is the mass of the Moon and d is the distance from the center of the spaceship to the center of the Moon. Since the spaceship is on the Moon's surface, d is equal to the radius of the Moon.

We've defined our unit for m_2 as "Earth-masses" and for d as "Earth-radii," and we will stick to that. The Moon is only 0.0123 (about ⅛₁) as massive as the Earth and its radius is only 0.273 (a little over ¼) that of the Earth.

The Moon's mass is therefore 0.0123 "Earth-masses" and its radius 0.273 "Earth-radii" so that Equation 2 becomes:

$$f = 0.0123/0.273^2 = 0.164 \qquad \text{(Equation 5)}$$

TABLE 1 Earth's gravitational force in relation to distance

Distance to surface of earth IN MILES	Distance to center of earth		Gravitational attraction
	IN MILES	IN EARTH-RADII	
0 (sea-level)	3,950	1.000	1.000
50 (top of the stratosphere)	4,000	1.012	0.975
150	4,100	1.040	0.924
250	4,200	1.063	0.884
1,000	4,950	1.253	0.636
2,000	5,950	1.506	0.442
4,000	7,950	2.015	0.247
10,000	13,950	3.53	0.081
20,000	23,950	6.06	0.027
50,000	53,950	13.62	0.0054
100,000	103,950	26.3	0.0014
250,000 (moon-apogee)	253,950	64.2	0.00024
25,000,000 (closest approach of Venus)	25,003,950	6300	0.000000025

This means that whatever the spaceship weighs on the surface of the Earth as the result of the force of Earth's gravitational attraction, it weighs 0.164 times that (roughly ⅙) on the surface of the Moon, as the result of the Moon's (lesser) gravitational attraction. By the same reasoning, this ratio of weight would hold true for any object at all.

Given the mass and radius of any body, the value of the surface gravity of that body can be calculated in the same way. The surface gravity of various bodies in the Solar System is presented in Table 2 by way of example.

Notice that Jupiter and Saturn are not perfect spheres. Both are noticeably flattened at the poles. Saturn is the least spherical of the planets, there being a 12 per cent difference between the polar radius and the equatorial radius. For Jupiter, there is a 7.5 per cent difference. In both cases, since d varies with latitude, so does surface gravity, being least at the equator and highest at the pole. (The equatorial gravity is further decreased by the centrifugal force of the planet's spin, but I've ignored that here. Enough is enough.)

The fact that Saturn, which is so much more massive than Earth, has a surface gravity only slightly higher is not mysterious. Saturn is only ⅛ as dense as Earth and is correspondingly more voluminous than it would be if it were made of Earth-type material. The effect of the abnormally large radius for Saturn's mass (as compared with Earth) is to lower the surface gravity because of increased distance between Saturn's center and an object on its surface by just about as much as Saturn's increased mass (over Earth) raises it.

Surface gravities of Saturn and Earth may be approximately equal but this is illusory, in a way. Look at it this way—

A spaceship on a planetary surface is at varying distance from that planet's center, since planets come in different sizes. Suppose, though, that a spaceship is 230,000 miles from Earth's center at one time and 230,000 miles from Saturn's center at another.

When it is 230,000 miles from Earth's center it is about 226,000 miles above its surface. At 230,000 miles from Saturn's center, it is only 192,000 miles above its surface, Saturn being the larger body. However, in considering gravitational force, as I have pointed out, it is distance from the center that counts.

TABLE 2   Some Surface Gravities in the Solar System

Astronomical Body	Mass (IN EARTH-MASSES)	Radius (IN EARTH-RADII)	Surface gravity
Jupiter (pole)	318.	10.5	2.88
Jupiter (equator)	318.	11.2	2.54
Neptune	17.3	3.4	1.50
Saturn (pole)	95.2	8.5	1.32
Saturn (equator)	95.2	9.5	1.05
Uranus	14.5	3.7	1.05
Earth	1.0	1.0	1.00
Venus	0.82	0.96	0.89
Mars	0.11	0.525	0.40
Mercury	0.054	0.380	0.27
Ganymede	0.026	0.395	0.17
Moon	0.0123	0.273	0.16

In such a case, with d equal in the two situations, only m_2 (see Equation 4) remains to vary the result. Earth's mass is, of course, equal to 1 "Earth-mass." Saturn's mass is 95.2 "Earth-masses." Therefore, the gravitational force gripping the spaceship in the neighborhood of Saturn is always 95.2 times that gripping it at an equal distance from Earth.

This can be shown in the behavior of two satellites that happen to be at this distance from Earth and Saturn. The Moon is at an average distance of 239,000 miles from Earth's center, while Saturn's satellite Dione is about 230,000 miles from Saturn's center. Each travels just about 1,500,000 miles in completing its circuit about its primary.

The greater the force of gravitational attraction upon a satellite, the faster must that satellite move to work up enough centrifugal force to keep in its orbit against its planet's pull. The Moon can manage this by traveling at a rate of 2200 miles an hour and completing its revolution in a leisurely 27.32 days. Dione, however, must race along at just ten times that speed to stay in orbit. Its period of revolution is only 2.74 days.

That, and not the surface gravity figures, is a measure of the force a spaceship would be fighting if it were maneuvering in the neighborhood of Saturn.

Nevertheless, however great the gravitational force exerted by a planet, and however close to it a spaceship may be, it remains possible for the spaceship (and the people on it) to be weightless. And this does *not* mean that the force of gravity has been suspended.

Gravity is a force and a force is defined as something that can accelerate a mass. That, so to speak, is gravity's main job. It is what it is doing constantly all over the Universe.

We ourselves happen to be most used to gravitational force in its manifestation as the sensation of weight. Actually this type of manifestation occurs only in a special case: where a body is prevented from responding to gravitational force by accelerated motion. (Accelerated motion, by the way, is motion that is continually changing either in velocity or in direction or both.)

The most common way in which accelerated motion can be prevented is by having the two bodies between which the gravitational force exists (i.e. a spaceship and Earth) in contact so that neither can move with respect to the other under the pull of gravitational force alone. You and I are almost always in contact with Earth and it is for that reason that we learn to think of gravity as primarily concerned with weight.

Yet we live with the acceleration too. Hold a book at arm level and let go. At once gravitational force expresses itself in terms of acceleration. The book accelerates in the direction of Earth's center and keeps on until the surface of the planet intercepts it and it can move no more.

The Moon, as it moves about the Earth, is undergoing accelerated motion since, moving in an ellipse as it does, it is continually changing direction, turning a full 360° in 27.32 days. (It also continually changes velocity to a comparatively minor extent.) Dione, under the whip of a stronger gravitational force, is more strongly accelerated, changing direction more quickly and turning 360°, as I have said, in only 2.74 days.

Whenever a body like a book or a satellite is responding to gravitational force by unrestricted accelerated motion, it is said to be in "free fall." The word "unrestricted" in the previous sentence is a bow in the direction of air resistance. A book falling from your hand ought to be moving through a vacuum to be in true free fall.

An object moving in response to gravitational force, with another constant (i.e. non-accelerated) motion superimposed, is still in free fall. A missile, with its charge expended, moving in a direction more or less opposed to that induced by gravity; or a satellite (artificial variety), with its rocket stages gone, and with a component motion perpendicular to that imposed by gravity—both are still in free fall.

An object which is completely in free fall is responding to gravity all it can; it has no response left over, so to speak, to be manifested as weight. An object in free fall is therefore weightless. A Cosmonaut orbiting the Earth in a satellite remains weightless as long as he stays in orbit. Gherman Titov remained weightless in this manner for a full day. For that matter, if the cable of an elevator broke and it fell freely with unfortunate you inside, you would be as weightless for a few seconds (barring air-resistance effects) as any man in orbit in outer space.

If you were falling at an acceleration *greater* than that imposed by gravity (as in an airplane power dive) you would feel "negative weight." Within such a power-diving plane, you would fall upward at increasing speed (relative to the plane) unless you were strapped into your seat. This is one kind of "anti-gravity" which may not be useful but which is at least completely valid.

In calculating the force of gravity at various distances from Earth and on the surface of various planets, I have compared these with the intensity of gravitational force on Earth's surface, which I arbitrarily set equal to 1.

But it is easy to measure the actual value of the gravitational force at Earth's surface. Since forces are measured by the accelerations they induce, it is only necessary to measure the acceleration of a body dropping, let us say, from the top of the Empire State Building to the ground under the influence of gravity. It turns out that this acceleration and, therefore, the value of the gravitational force (at the equator, at sea level, and corrected for the effects of air resistance) is 980.665 centimeters per second per second, or, in more familiar units, 31.6 feet per second per second.

This means that if an office safe is raised to a height of 5000 feet above the Earth's surface and released, it would fall at the rate of 31.6 feet/sec after one second, twice that (63.2 feet/sec) after two seconds, three times that (94.8 feet/sec) after three seconds, and so on, its rate of fall increasing smoothly with time. (Here and elsewhere in this chapter, I am ignoring the effects of air resistance, which is a subversive influence and a nuisance.)

The equation relating the distance (s) through which a

ody falls during a time (t) under gravitational acceleration
(g) is:

$$s = \tfrac{1}{2}gt^2 \qquad \text{(Equation 6)}$$

The value of g is, of course, 31.6, and if a body is falling
from 5000 feet above Earth's surface, s is 5000. By substituting these figures into Equation 6, it can be solved for t.
It turns out that it will take our office safe 17.8 seconds of
fall before it splashes into Earth's surface. At the time of
contact, it will be moving 17.8 × 31.6 or 562.5 feet/sec (or
0.106 miles/sec).

(It does not, by the way, matter, whether we use a golf
ball or an office safe as the falling object. The inertia of an
object varies directly with its mass, which means it takes
twice the force to accelerate a two-pound weight at a certain
rate as it does to accelerate a one-pound weight. But gravitational force also varies with the mass of the falling object.
A two-pound weight is attracted to Earth with twice the
force of a one-pound weight. Generalizing this, you can see
that the end result is that all objects, whatever their mass,
experience the same acceleration in a given gravitational
field. The effect of air resistance on light objects, such as
feathers and leaves, obscures this fact and misled Aristotle—
who thought a two-pound weight fell with twice the acceleration of a one-pound weight—and all who followed him down
to the time of Galileo.)

The figures on fall under gravity are true in reverse also.
If a cannonball is shot directly upward against Earth's gravity, at a velocity of 0.106 miles/sec as it leaves the cannon's
mouth, it will travel upward (slowing constantly) for 17.8
seconds and reach a height of 5000 feet before coming to a
halt and beginning to fall back.

If our original office safe were raised to a height of 20,000
feet instead of 5000, the time of fall would then be 35.6 seconds and the final velocity is 0.212 miles/sec. And if the cannonball were shot upward at an original velocity of 0.212
miles/sec—but you can see that without my telling you.

It follows, generally, from Equation 6, that the time of
fall and the final velocity of a falling object, vary as the
square root of the distance of fall, assuming a given constant value of g. It would seem then that the final velocity
at contact of office safe and Earth could be as high as you
are to make it—by setting the safe to falling from a greater
and greater height above the surface.

But there's a catch. I said we must assume "a given constant value of g," and that is exactly what we can't do.

The value of g varies with distance from the Earth's center, as I explained earlier. In lifting an office safe, or a golf ball, 5000 or even 20,000 feet above Earth's surface, the distance from Earth's center is not significantly changed, and you can work your calculations as though g were constant.

But suppose you were to release your object 3950 miles above the surface of the Earth. Up there, the value of g is only 0.25 what it is on the surface and the acceleration imposed upon a falling body is likewise only 0.25 what it is here on the surface.

To be sure, the value of g increases as the object drops and is a full 1 g by the time it is at the collision point. Nevertheless it takes longer for the object to complete its drop than it would have if the value of g were 1 all the way down, and it doesn't hit at as high a velocity as it would if the value of g were 1 all the way down.

Every additional thousand miles upward from Earth's surface adds less and less to the final velocity. The result is a converging series where an infinite number of smaller and smaller terms add up to a finite sum. This finite sum, in the case of objects falling toward Earth, is 6.98 miles/sec. This means that if an office safe, or anything else, were to fall from any distance, however great, its final velocity as it struck Earth would never exceed 6.98 miles/sec.

This figure might be called the "maximum final falling velocity," but it isn't. People prefer to look at it in reverse. If a cannonball, a spaceship, or anything else were fired directly upward at a velocity of 6.98 miles/sec (or more), it would continue moving outward indefinitely, if there were no interference from extraneous gravitational fields. (Since a fall even from an infinite distance could not create a final speed of more than 6.98 miles/sec, then the reverse follows: An initial speed of 6.98 miles/sec or more could never be reduced to zero by Earth's gravitation, even if the object traveled forever.)

An object hurled out in this fashion would never return to Earth. It will not have escaped from the influence of the Earth's gravitational field (which will be slowing it constantly) but it will have escaped Earth itself.

So the velocity of 6.98 miles a second is the "escape velocity" for Earth.

The value of the escape velocity varies with the mass of

the attracting body and the distance from its center as follows:

$$v = 6.98 \sqrt{m/d} \qquad \text{(Equation 7)}$$

where v is the escape velocity, m is the mass of the attracting body in "Earth-masses" and d the distance to the center of the attracting body in "Earth-radii." The factor 6.98 allows the escape velocity to come out in miles per second.

The Moon, for instance, has a mass equal to 0.0123 "Earth-masses" and, at its surface, the distance from its center is 0.273 "Earth-radii." The escape velocity from the Moon's surface is therefore $6.98 \times \sqrt{0.0123/0.273}$, or 1.49 miles/sec.

The escape velocities at the surface of any body in the Solar System can be similarly calculated and the results are presented in Table 3.

One caution: Escape velocity is required for escape from a planet only where unpowered (i.e. "ballistic") flight is concerned. If you are in a spaceship under constant power, you can move any finite distance from Earth at any velocity below escape velocity but above zero, provided you have fuel enough. (In the same way, you cannot jump to a second story window at a bound unless the initial thrust of your leg muscles against the ground is great enough—which is more than you can manage—but you can nevertheless walk up two flights of stairs as slowly as you please.)

And yet escape from Earth may be not entirely escape, either. I said earlier that an object hurled from Earth at more than escape velocity would move outward forever "if there were no interference from extraneous gravitational fields."

TABLE 3 Escape Velocities at the Surface of Bodies in the Solar System

Astronomical Body	Mass (EARTH-MASSES)	Radius (EARTH-RADII)	Escape Velocity (MILES PER SECOND)
Jupiter (pole)	318.	10.5	38.4
Jupiter (equator)	318.	11.2	37.3
Saturn (pole)	95.2	8.5	23.4
Saturn (equator)	95.2	9.5	22.1
Neptune	17.3	3.4	15.8

Uranus	14.5	3.7	13.9
Earth	1.0	1.0	6.98
Venus	0.82	0.96	6.46
Mars	0.11	0.525	3.20
Mercury	0.054	0.380	2.64
Ganymede	0.026	0.395	1.80
Moon	0.0123	0.273	1.49

But, of course, there *is* such interference. Consider the Sun, for instance, which so far we haven't done.

The Sun has a mass that is equal to 330,000 "Earth-masses" and a radius equal to 109 "Earth-radii." Using Equation 7, the escape velocity from the Sun's surface turns out to be a tidy 385 miles/sec.

From Earth, however, the distance to the Sun's center is about 23,000 "Earth-radii." Substituting that figure for d in Equation 7, and leaving m at 330,000 "Earth-masses," it turns out that the escape velocity from the Sun at Earth's distance is 26.4 miles/sec.

This is four times as high as the escape velocity from Earth itself. In other words, a missile shot out from Earth and attaining a velocity of 6.98 miles/sec by the time the rocket thrust is expended, may be free of the Earth, but *it is not free of the Sun*. It will not recede forever after all, but will take up an orbit about the Sun.

To escape from the Solar System altogether, a speed of 26.4 miles/sec must be attained in ballistic flight. To be sure, in powered flight, we don't have to attain escape velocity; we can just keep the engines going. However, the escape velocity is a measure of the amount of energy we must use to break the gravitational chains in any fashion. So you see, it is the Solar prison bars that block our way to the stars far more than Earth's puny fence.

The only consolation is that, for the moment, the Moon and the planets are enough of a challenge. The stars can wait.

5 Of Capture and Escape

Since January 2, 1959, the Soviet Union and the United States have sent up a number of missiles which were notable for three things:

(1) They reached and passed the orbit of the Moon.

(2) They were not captured by the Moon; that is, they did not take up a closed orbit about the Moon alone.

(3) They took up a closed orbit about the Sun and became artificial planets.

I'd like to consider each of these points in turn.

First, what does it take to reach the orbit of the Moon by means of a ballistic missile? (A ballistic missile is any projectile which receives an initial impulse of some sort and thereafter moves under the influence of gravitational forces only.)

If such a missile is fired straight up (i.e. directly away from Earth's center) the maximum height it will reach will depend (a) on the strength of the initial impulse upward and (b) the strength of Earth's gravitational pull downward.

Naturally, the greater the initial impulse upward, the greater the height reached. You might expect that doubling the initial impulse will double the height reached, but that is too pessimistic. It would be so if the gravitational force remained constant all the way up, but it does not. The higher the missile reaches, the weaker the gravitational drag upon it. The second half of its climb meets less resistance therefore and is correspondingly extended.

Consequently, doubling the initial impulse *more* than doubles the maximum height reached, and the more you increase the initial impulse, the more drastically do you increase the maximum height reached.

Table 1 gives the maximum height attained for various

initial velocities of the missile. The initial velocity is a measure of the strength of the push given the missile. (Naturally, there are complicating factors. There is air resistance; there is the fact that the push of the rocket motors isn't administered instantaneously, but is spread over several minutes, and so on. Since we're all friends here, I'm taking the privilege of ignoring such matters and leaving them to the missile engineers, who are most welcome to them.)

Notice how quickly the maximum height increases, especially at speeds higher than 6 miles a second, or, if you prefer, 21,600 miles an hour. (I have always had a liking for the use of "miles per second" as the unit for high velocities, but to a nation of automobile drivers "miles per hour" seems more natural. Besides, newspapers and allied information-mongers use "miles per hour" exclusively, perhaps because larger and flashier numbers are involved. So I'll use both units throughout. I just wish to warn you, though, that 21,600 miles an hour may sound flashier than 6 miles a second, but the two are entirely equivalent.)

A missile leaving Earth with an initial velocity of 6.92 miles a second (24,912 miles an hour) will reach a height of 220,000 miles before coming to a halt and beginning to fall back. This is just about the distance of the Moon at its closest approach ("perigee") to the Earth.

If, however, the missile leaves Earth at a velocity of 6.90 miles a second (24,840 miles an hour), it falls 50,000 miles short of the Moon. A difference of 0.02 miles a second (72 miles an hour) to begin with means a 50,000 mile discrepancy to end with.

TABLE 1

Initial Velocity of Missile		Maximum Height above Earth's Surface (MILES)
(MILES PER SECOND)	(MILES PER HOUR)	
1	3,600	80
2	7,200	350
3	10,800	900
4	14,400	1,940
5	18,000	4,180
5.5	19,800	6,450
6.0	21,600	11,100
6.5	23,400	25,800

6.6	23,760	34,300
6.7	24,120	46,300
6.8	24,480	73,600
6.85	24,660	102,800
6.90	24,840	.170,000
6.92	24,910	221,000
6.95	25,020	454,000
6.98	25,130	∞

It is for this reason that when one of our early Moon-probes only reached a third of the way to the Moon, it did *not* mean we had only attained a third of the necessary velocity. Actually, we had attained over 98 per cent of the necessary velocity. It's just that the last per cent or so is what carries the missile the remaining two thirds of the way to the Moon.

To go back to Table 1, a missile leaving Earth at a velocity of 6.98 miles a second (25,130 miles an hour—or something like 216 miles an hour faster than is required to reach the Moon's orbit) has no maximum height. If you like, its maximum height is infinite, symbolized as ∞ in the table. Such a missile would move away from Earth forever, assuming there is no interference from gravitational fields of other bodies. The velocity of 6.98 miles a second (25,130 miles an hour) is therefore the "escape velocity" from Earth's surface.

Imagine a missile that has left the Earth's surface at just the escape velocity. As it travels away from the Earth, its velocity decreases inversely as the square root of its distance from Earth's center. (When the distance has been multiplied by 4, the velocity has been decreased by 2.) The result is shown in Table 2.

Earth's gravitational pull is constantly decreasing the missile's velocity, but with increasing distance, the pull loses power and decreases the velocity at a slower and slower rate. The velocity therefore gets closer and closer to zero as the missile recedes from Earth, but never quite gets to zero.

If the missile had left at less than the escape velocity, Earth's gravity would have managed to bring the missile's velocity to zero at some finite distance and the missile would then fall back. If the missile leaves at a speed greater than the escape velocity, its velocity decreases and decreases with distance but never falls below a certain velocity, greater than zero, however far it travels. (All this assumes the presence of no other gravitational fields in the Universe, gumming up the works.)

Let's put it another way. A missile leaving Earth at a velocity less than the escape velocity follows an elliptical orbit. An ellipse is a closed curve, so that the missile does not depart more than a certain distance from the Earth. If the elliptical orbit happens to intersect Earth's surface, the missile crashes its first time round, as our first Moon-probes did. If the elliptical orbit does not intersect the Earth's surface, artificial satellites are the result.

TABLE 2

Distance from the Center of the Earth	Velocity of Missile Fired at Escape Velocity	
(MILES)	(MILES PER SECOND)	(MILES PER HOUR)
4,000 (Earth's surface)	6.98	25,130
8,000	4.93	17,800
12,000	4.04	14,500
16,000	3.49	12,550
20,000	3.12	11,210
40,000	2.21	7,950
80,000	1.56	5,620
120,000	1.27	4,570
160,000	1.10	3,960
221,000 (Moon at perigee)	0.95	3,410
253,000 (Moon at apogee)	0.88	3,160
400,000	0.70	2,510
1,000,000	0.44	1,580
∞	0.00	0

A missile leaving Earth at a velocity just equal to escape velocity takes up a parabolic orbit. A parabola is an open curve that never turns back on itself. Consequently, any object leaving Earth on a parabolic orbit never returns, barring the interference of the gravitational fields of other heavenly bodies.

If a missile leaves Earth at more than escape velocity, it follows a hyperbolic orbit. A hyperbola is also an open curve —even more open than a parabola, in a manner of speaking —so again the missile never returns.

Returning now to Table 2 (this gets complicated but I'm slowly building up a line of argument which, I hope, I can put to good use) I want to point out a special significance of the "velocity" column. The velocity of the missile which began at escape velocity remains at escape velocity throughout!

To be sure, the actual velocity of the missile is continually decreasing as its distance from Earth increases; but so does the escape velocity. And the escape velocity keeps pace throughout; for it, too, varies inversely as the square root of the distance from Earth.

Suppose you were to start from scratch at a distance 8000 miles from Earth's center, which is just about 4000 miles above Earth's surface. (Imagine, in other words, that you were on top of a mountain—a mythical one—4000 miles high.) Up there Earth's gravitational pull would be only one fourth what it is at sea level. There would be that much less drag on the missile and a smaller initial velocity would suffice to kick it into a parabolic orbit. To be exact, 4.93 miles a second (17,800 miles an hour) would suffice.

And from a mountain 80,000 miles high, an initial velocity of 1.56 miles a second (5620 miles an hour) would suffice. And from a mountain 1,000,000 miles high, 0.44 miles a second (1580 miles an hour) would suffice.

But at no finite distance from Earth, however great, would escape velocity actually be zero. At any finite distance, an object completely at rest with respect to Earth, would start moving toward the Earth in response to its gravitational pull —provided no other gravitational field interferes. To prevent the object from falling to Earth, some definite opposing push is needed; perhaps an infinitesimally small one if the distance is great, but some push is needed.

All this holds true for a missile (or a meteor) passing close by Earth from some outer-space starting point.

Suppose a meteor passed Earth at a distance of 120,000 miles from its center and had a velocity (with respect to Earth) of less than 1.27 miles a second (4570 miles an hour). Since the meteor's velocity is less than the escape velocity at its point of approach, it is forced into an elliptical orbit about the Earth. It is captured.

If its velocity were exactly 1.27 miles a second (4570 miles an hour) it would take up a parabolic orbit; if its velocity were greater it would take up a hyperbolic orbit. In both these latter cases, its direction of travel would be changed and it would curve about Earth more or less sharply.

But in neither case would it be captured. It would go shooting off into space never to return.

Of course, both parabolic and hyperbolic orbits travel about the *center* of the Earth as a focus. If the meteor is aimed in such a fashion that its new orbit will pass within 4000 miles of the Earth's center, it will intersect Earth's surface. The meteor will then enter our atmosphere and flame to death. However, hitting the Earth is not the same as being captured by the Earth.

Since escape velocity increases with decreasing distance from Earth, a meteor is more likely to be captured if it passes close to the Earth, than if it passes at a distance. A meteor traveling at a velocity of 3.12 miles a second (11,210 miles per hour) relative to the Earth, will be captured if it passes Earth at a distance of less than 20,000 miles, but not if it passes Earth at a distance greater than that. Below 20,000 miles its velocity is less than escape velocity; above, it is higher than escape velocity.

The more massive a planet is, the higher its escape velocity at all distances, and the more likely it is to capture invading meteors and planetoids. Jupiter, for instance, with a mass 318 times that of Earth has an escape velocity at its surface of 37.3 miles per second (134,000 miles an hour). Since Jupiter's surface is some 40,000 miles from its center, the comparable escape velocity in the case of Earth is only 2.21 miles a second (7950 miles an hour). At a distance of 1,000,000 miles from Jupiter's center, the escape velocity is 13.2 miles a second (47,500 miles an hour) as compared to 0.44 miles a second (1580 miles an hour) for a comparable distance from Earth.

It is not surprising then that the seven outermost of Jupiter's twelve satellites are generally considered to be captured planetoids. But if a more massive planet is a more efficient capturer of wandering objects, a less massive astronomical body should be a less efficient capturer. That brings us to the Moon, which is only $\frac{1}{81}$ as massive as the Earth and should therefore be a very poor capturer of meteors and assorted debris such as missiles.

The escape velocity from the Moon's surface is a mere 1.49 miles a second (5360 miles an hour) and this falls off, in the usual way, in inverse ratio to the square root of the distance from the Moon's center. The escape velocity at various distances from the Moon is given in Table 3.

To be captured by the Moon, a missile must pass the

Moon at a velocity less than the escape velocity at that distance. What's more, the velocity involved is the velocity relative to the Moon, not relative to the Earth.

The Moon, you see, is itself moving at a velocity of about 0.64 miles a second (2300 miles an hour) with respect to the Earth. Suppose, then, a missile shot from Earth at 6.92 miles a second (24,912 miles an hour) just makes it to the Moon's orbit and hangs momentarily suspended at zero velocity (with respect to the Earth) at a distance of 4500 miles from the Moon's surface (5500 from its center).

The Moon, however, is retreating from it, or advancing toward it, or passing to one side of it (depending on the exact position of the missile with respect to the Moon) at 0.64 miles a second (2300 miles an hour), so that is the missile's velocity *relative to the Moon*. This velocity is just a bit over the Moon's escape velocity at the distance of 5500 miles from its center.

If the missile had been fired with a greater initial velocity, so that it was still moving at some velocity or other when it reached the Moon's orbit, its velocity relative to the Moon would be greater still.

It follows then that any missile that misses the Moon's center by 5500 miles or more cannot be captured by the Moon and will not move into an orbit about the Moon, no matter how slowly the missile is going. The respective motions may be such that the missile may *hit* the Moon, as did the Soviet Union's Lunik II, but that's another thing. It may hit the Moon but it won't be captured by the Moon in the sense that it will go into a closed orbit about it.

TABLE 3

Distance from the Center of the Moon	Velocity of Missile Fired From Moon at Escape Velocity	
(MILES)	(MILES PER SECOND)	(MILES PER HOUR)
1,000 (Moon's surface)	1.49	5,360
1,500	1.21	4,360
2,000	1.06	3,820
2,500	0.94	3,380
3,000	0.86	3,100
3,500	0.80	2,880
4,000	0.74	2,560
4,500	0.70	2,520

5,000	0.66	2,375
5,500	0.63	2,270
∞	0.00	0

A missile fired from Earth at escape velocity will pass the Moon (at perigee) at 0.95 miles a second (3410 miles an hour). Thanks to the Moon's own motion, which will be roughly at right angles to that of the missile, the missile's velocity with respect to the Moon will be 1.15 miles a second (4140 miles an hour). This is the Moon's escape velocity at a distance of about 1600 miles from its center. Such a missile would therefore have to come within 600 miles of the Moon's surface before it can be captured and go into an orbit about the Moon.

A missile fired from Earth at 7.37 miles per second (26,500 miles an hour) will pass the Moon at a velocity of 1.34 miles per second (4820 miles an hour) with respect to the Earth, but a velocity of 1.49 miles per second (5360 miles an hour) with respect to the Moon. This is the Moon's escape velocity at its surface. A missile fired from Earth at this velocity or above cannot be captured by the Moon, no matter how close to the Moon it passes, not even if it grazes its surface. (I repeat, it can *hit* the Moon, but again I repeat, that's a different thing.)

So the limits for success are narrow indeed. A missile must be fired at a velocity of at least 6.92 miles per second (24,910 miles an hour) or it won't reach the Moon; and it must be fired at a velocity of less than 7.37 miles per second (26,500 miles an hour) or it can't be captured by the Moon. And even within that narrow range of velocities, capture by the Moon is only possible if the missile passes quite close to the Moon. A miss of not more than 4500 miles from the Moon's surface is the maximum, and this leeway rapidly decreases as you approach the upper limit of the permissible range.

In fact, ballistic missiles are so hard to place into an orbit about the Moon that I wonder if it's even sensible to try. It might be better to make the missile non-ballistic. That is, to supply a final delayed rocket blast which could be set off by radio at such a time and in such a direction as to decrease the velocity of the missile relative to the Moon and make it capturable.

This brings us to the final point I raised at the beginning of the article, the question of orbiting about the Sun.

As I pointed out in the chapter "Catching Up With Newton," the escape velocity from the Sun, even way out here at Earth's orbit, 93 million miles from the Sun, is still 26.4 miles per second (95,040 miles an hour). I left it at that point then, but let's carry it further now.

The figure 26.4 miles a second (95,040 miles an hour) refers, of course, to velocity relative to the Sun. If the Earth were at rest with respect to the Sun, we would have to fire a missile at that initial velocity to free it of the Sun's grip. However, the Earth is *not* at rest with respect to the Sun, but travels in an orbit about the Sun at a velocity of 18.5 miles a second (66,600 miles an hour).

Suppose, then, we were to fire a missile in the direction of the Earth's motion. It would already be traveling 18.5 miles a second (66,600 miles an hour) with respect to the Sun before it started. Giving it additional velocity would raise the figure (like flying an airplane downwind). A velocity, relative to the Earth, of 7.9 miles a second (28,440 miles an hour) would just suffice to raise the missile's velocity to the point where it could escape the Solar System altogether, provided it didn't hit something on the way.

This is the most economical way of freeing a missile from the grip of both Earth and Sun.

If a missile were fired at right angles to Earth's motion, either directly toward or away from the Sun, it would receive some but not all the benefit of Earth's motion (like an airplane flying cross-wind). The missile would have to be fired at an initial velocity of 18.8 miles a second (67,680 miles an hour) to attain to Solar System escape.

If it were fired in the direction opposite to the Earth's motion, Earth's motion would then not be helping but hindering. The missile would require the full initial velocity of escape from the Sun plus enough more to neutralize the Earth's motion (like an airplane flying upwind.) For a missile so fired to escape would require an initial velocity of 44.9 miles a second (161,600 miles an hour).

The first successful Moon-probe was fired at a time when the Moon was in "last quarter." At this time, the Moon is directly ahead of the Earth in their path around the Sun, so the probe was fired in the direction of Earth's motion. Nevertheless, if we remember that the probable initial velocity of the missile might have been as high as 7.5 miles a second (27,000 miles an hour) this is still insufficient to allow escape from the Sun, and the missile remained in orbit about the Sun.

To be sure, it has a higher velocity than the Earth has so that its orbit bellies out into the space between Earth and Mars. (Since the missile's velocity is higher than Earth's, it makes a slightly more effective attempt, so to speak, to get away from the Sun, and it gets halfway to Mars before the Sun pulls it back.) As a result, the missile's year is 15 months long, rather than 12 months long as is our Earth's.

The two orbits cross, however, and it is conceivable that someday both missile and Earth may be at the crossing point simultaneously, in which case the missile will finally come home.

One last question. Was there any chance that a missile such as Lunik I or Pioneer IV might have fallen into the Sun?

Well, let's see what's required to hit the Sun. Suppose you aimed a missile directly at the Sun. It would travel toward the Sun, yes, but at the same time it would retain Earth's motion of 18.5 miles a second (66,600 miles an hour) in a direction at right angles to its aimed line of motion at the Sun. Its over-all motion would be a combination of both component motions. Earth's sidewise motion would therefore carry the missile around the Sun in an elliptical orbit, if its initial velocity with respect to Earth were less than 18.8 miles a second (67,680 miles an hour)—this being the Solar escape velocity for a missile fired at right angles to Earth's motion.

If the missile were fired at exactly the escape velocity, the component due to Earth's motion would carry the missile about the Sun in a parabolic orbit; if it were greater than escape velocity it would go about the Sun in a hyperbolic orbit.

The greater the velocity in the direction of the Sun, the flatter the hyperbola and the closer it would approach the center of the Sun at its closest approach. If you aimed at the center of the Sun, no velocity short of the infinite would enable you to hit the center, thanks to the sidewise component of motion.

Of course, why aim at the Sun's center? Why not aim to one side of it, allowing Earth's motion to bring the missile to the Sun; instead of aiming at it and allowing Earth's motion to carry the missile past it. (This is like allowing for the wind when you aim a gun.)

The most economical way to neutralize Earth's motion is to shoot the missile in a direction directly opposite to that motion. If the missile is then fired at a velocity of just 18.5

miles a second (66,600 miles an hour), Earth's motion with respect to the Sun is neutralized. The missile is, in fact, then at rest with respect to the Sun, and it will proceed to fall into the Sun under the inexorable pull of Solar gravity.

If a missile is fired in this opposite direction to Earth's motion (that is, at the Moon at "first quarter") at less than this speed, its motion with respect to the Sun is still less than that of the Earth. It would not fall into the Sun, but it would approach it more closely than does Earth, its orbit moving in toward that of Venus. We would then have a Venus-probe as in the case of Pioneer V.

There's a lesson here. A spaceflight to Mars must start off in the direction of Earth's motion, while one to Venus must start off in the direction opposite Earth's motion—at least if we want to make economical use of the motion we are already blessed with from birth.

PART II / THE SOLAR SYSTEM

6 Catskills in the Sky

Once I received, as a gift, a record entitled "Space Songs." It was intended for my children and so I called them both to my record player and we listened. They liked it, but as it happened, I liked it even more than they did. Realizing, unlike Sir Philip Sidney, that my need was greater than theirs, I quietly added it to my own record collection and have listened to it periodically ever since.

Anyway, to get to the point, one of the songs on the record is entitled "Why Go Up There?" and the words are:

> Why do we all want to be
> >   up there—up there?
> What is there to do or see
> >   up there—up there?
> Outer space
> Is a place
> Where we'll trace
> >   the future.
> There's a lot
> Of who knows what
> >   away—up there.

As you see, the reasons given to go up there are a bit vague, and I intend to correct that now. Let's consider some of the "who knows what" that will serve as strong inducements for the average man or woman to travel long distances away from the Earth.

Imagine a society in which spaceflight is routine, and no more difficult or remarkable than airflight is now, or trainflight was in the nineteenth century, or coachflight in the eighteenth century. Well, then, why should anyone want to go to the Moon?

For the same reason, it seems to me, that people nowadays want to go to Switzerland or Pakistan or Brazil; to see new sights, do new things and, in general, feel the stimulation of sensations never before experienced.

Presumably there will be a time when the schoolteacher from Dubuque and the curious young man from Düsseldorf will carry their cameras along on some Cook's tour of the Moon, just to see the Moon and send back appropriate picture postcards (by rocket mail, of course) to their stodgy stay-at-home friends.

Naturally there are many wild and grand things on the Moon that are not to be seen or experienced on the Earth; the vast silences, the bright, unwinking stars, the slowly moving inferno of the Sun, the trackless dust and the craggy peaks and ring-shaped crater walls, lit in the soft light of Earth.

And of course, of all the unique sights the foremost would be that of the Earth itself. I imagine that a picture of the Earth suspended in the sky would be on at least three fourths of the picture postcards manufactured for the tourist trade, and if the Moon ever has a flag of its own, that flag will feature a white Earth on a black field.

The Earth as viewed from the Moon is far more impressive than the Moon as viewed from the Earth. The Earth's globe would be nearly four times the diameter of the Moon as seen by us now and it would have thirteen times the area. Furthermore, the Earth reflects much more light than the Moon does (thanks to the Earth's atmosphere) and there is no air blanket on the Moon to sop up any of that reflected light. So the Earth ends up some seventy times as bright as the Moon appears to us.

What's more, the Earth would be much more interesting to look at. It would go through the same phases at the same rate that the Moon does, but the terminator (the line between light and dark) would not be the sharp, uninteresting boundary it is on the Moon. Again thanks to the Earth's atmosphere, it would be a softly gradual darkening, a visible fading of day into night.

The continents and oceans would not be clearly visible through the Earth's cloudy and light-scattering atmosphere, but the globe would have a blue-white appearance arranged in misty bands (because of Earth's atmospheric circulation) parallel to the equator. There may be washes of deeper blue, of blue-green, of faint orange to mark ocean, fertile land, and desert.

In particular, the sight of Earth would be wonderful on those occasions when the Sun travels behind it and is hidden. (To us on Earth, such periods are "lunar eclipses.")

On such occasions, the Sun would approach the Earth from the east and the Earth would be visible only as a thin crescent, convex toward the Sun and probably lost in its glare. The Solar corona, which might also be lost in the glare of the Sun itself, would move behind the Earth's globe first. More and more of the corona would be hidden until the globe of the Sun was bitten into. It would take just about one hour for the Sun's globe to disappear completely behind the Earth after initial contact.

During that hour, all the tourists would undoubtedly be watching from beneath a transparent dome fitted with special filters to cut out ultraviolet and most of the visible light. With the complete disappearance of the Sun's globe, the filters would be removed and the spectacle would be visible in full clarity and glory.

The corona itself, pearly white, would come into full view, its streamers extending beyond the Earth on all sides. Between the corona and the black inner circle of the Earth would be a thin ring of orange fire! This would mark the sunlight, refracted ruddily through the Earth's atmosphere on all sides.

Undoubtedly, the Moon tours would feature special excursions to catch the eclipses and I can imagine the disappointment that would follow if climatic conditions on Earth were such that those sections of the atmosphere which happened to be exposed about the rim of our planet at the time of the eclipse were filled with clouds so that the ring of orange light did not come into view. (This actually happens sometimes, for although the Moon is usually a coppery color during total eclipse thanks to the light received from the orange ring of refracted sunlight, it does, on a few occasions black out entirely—no ring.) I can safely predict that some enterprising company will set up special "eclipse insurance" that, for an appropriate premium, would guarantee the return of the travel fee in case the ring doesn't show.

And of course, since the Earth can be seen only from the side of the Moon facing us, that side will be much more valuable to the concessionaires. Owning land on the other side of the Moon will be much like owning a mountain resort which is not on a lake. (Nevertheless I can see the advertising folders now, making the most of the Other Side: "Be lost in the wonders never before seen by man. The

Other Side, mysterious, haunting, hidden for long ages from all prying eyes, is now YOURS!")

But the Moon has more to offer than the sight of its skies. It has its low gravity. There is no question but that this would mean fun for the tourist. Twenty-foot high jumps, sixty-foot broad jumps, every man an athlete, whee-e-e-e-e.

Yet low gravity wouldn't be all honey and soda water either. Anyone who expected to stay on the Moon for any length of time would have to get used to new ways of manipulating objects. On the Earth we correlate weight and mass through a lifetime of practice. From the muscular effort it takes to lift a heavy medicine ball, we know in advance just about the oof it will take to catch one in the pit of the stomach.

On the Moon, weight (which is the measure of gravitational pull) goes down, but mass (an inalienable property of matter) does not change. The two no longer match. It is easier to pick up the medicine ball and therefore it may be natural to think it would take less of an oof to interpose the pit of the stomach, but it will not. The oof depends on the mass, not the weight, and until you learn to allow for that, you will be in trouble every time.

Again, you jump upward on the Moon in slow motion since the Moon's gravitational pull will slow you (as you go up) and speed you (as you go down) at only one sixth the acceleration that the Earth's gravitational pull would. If you jump with all your might, however, you leave the Moon's surface and eventually strike it again, with the same velocity with which you would leave and strike the Earth's. You will land with the usual momentum. If, therefore, you are deluded by the slowness of the jump into thinking you are going to land like a feather on the tippy-tippy-tip of your toe, and try to do so, you will very likely break your ankle.

Then nothing is so easy to get used to as luxury and once your muscles get the hang of low gravity, they will get to like it and learn to expend no more effort than necessary. They will probably weaken rapidly and turn to flab. No harm to this on the Moon but what about the day you land on Earth and find your muscles are protesting violently at the sudden sixfold increase in weight?

In fact, I'll get out on a limb and predict that when the Moon is colonized, people who expect to travel back to Earth now and then will have to undergo an established period of exercise under Earth-normal gravity to keep their muscles in tone. One practical way in which this might be done would be to maintain a large centrifuge which can be whirled to the point of inducing a centrifugal force equal to Earth's gravity.

I can see tourists being herded into the centrifuge in shifts each day with a grim, no-nonsense drillmaster in charge insisting on a full regimen of calisthenics. Of course, there would be the inevitable wise guy who succeeds in goofing off, and the price he pays in semi-collapse when he gets back to the Earth will be well-deserved.

Then there is the possibility that some human beings will deliberately want the low gravity as a permanent thing. When retirement age comes, old hearts that must struggle to pump a weight of blood against gravity and old muscles that must struggle to lift the weight of the body, will obviously be benefited by having some of that weight removed. The old (provided they can afford the price of a ticket and are strong enough to withstand the accelerational rigors of the journey) might well find decades added to their life, if they spend those decades on the Moon.

The decision to spend one's final decades on the Moon might well be irreversible, however. I don't see the aged being able to resume the remaining five sixths of their weight, once they have been relieved of it for any period of time. And yet some may, too late, regret their decision. Too late, they will long for home.

I suppose one could write a story about one of them, watching the globe of the Earth with his heart in his eyes, hanging about the fringe of tourist parties with hopeless yearning and, eventually, managing to stow away on a ship bound back for Earth. The acceleration would nearly kill him, of course, and he would be dying when discovered. But he would have one last look at Earth's green hills, one last breath of free air, even one last thankful sensation of the heavy drag of Earth's gravity before death.

We can look a bit further into the future, when flights to the Moon become so tame as to become *déclassé*, and of no more importance than a trip to the Catskills or Cape Cod. "My dear, no one, but no one, goes to the Moon any more. It's just *filled* with common people."

But what can we get elsewhere that we cannot get on either the Earth or the Moon? What can we get to attract the tourist trade? Of the two nearest planetary targets, Venus is perpetually cloud-covered and there is no way of predicting what its surface would be like.* We *can* predict that its sky will be a uniform and perpetual gray, and right away I denounce that as unbearably depressing. For that we can visit London.

Mars, on the other hand, has two moons in a clear sky—wow! Any number of romantic descriptions have been written, emphasizing how doubly stimulating it would be for a young couple to look at two moons instead of one.

Unfortunately, this is sheer moonshine. One of Mars's two moons is no moon at all in our sense of the word. I refer to Deimos, the outer of the two, which is nothing but a mountain on the loose. It is five miles in diameter and since it is 12,500 miles from Mars's surface, it shows no visible disk. It is a mere point of light just about as bright, when seen from Mars's surface, as Venus seems to us on Earth's surface.

Phobos is not much larger, being only about ten miles in diameter. However, it circles only 3600 miles above the surface of Mars, so when it is directly overhead it is, despite its small size, about a third of the diameter of the Moon as seen from the Earth. When it is close to the horizon, it is further away by the radius of Mars itself so that its apparent diameter is cut almost in half.

At zenith it would be only 1/20 as bright as the Moon, and only 1/60 as bright when it is near the horizon. Because of its small size, Phobos might well be irregular in shape and it might be amusing to watch a craggy moon instead of an uninterestingly smooth one.

There is another point about Phobos that would interest the tourist. Small and dim it may be, but Great Galaxy, it *moves!* It revolves about Mars in 7 hours and 40 minutes. This is faster than Mars rotates about its own axis (24½ hours). Phobos therefore overtakes the Martian surface and rises in the West and sets in the East.

To an observer on Mars, Phobos would stream from western horizon to eastern in about 5½ hours. It would move quickly enough to have its motion visible to the naked eye.

* Our Venus probe of late 1962, however, confirmed earlier suspicions, arising from Venus's radiation of microwaves, that the planet's surface has a temperature of from 600 to 800° C. This means that the surface of Venus is hot enough to glow a dim red. That is even more depressing than a gray sky. I. A.

And, to top it off, it would change phases as it traveled, going through more than half the cycle during the period in which it was above the horizon.

Certainly this would at least partially compensate for its smallness and dimness compared with our Moon, and the Martian vacation folders would undoubtedly go heavily to town on the subject of Phobos's motion, with pictures that would probably shamelessly exaggerate its size. Of course the real estate owners on Mars would have to be careful. Phobos is so near Mars's surface that the bulge of Mars's globe cuts off the view of the moon from any observer near the Martian poles. A tourist haven on the planet must not be located too far north or south if a view of Phobos is desired.

(The one interesting object in the skies of Mars apart from the two moons would be Earth itself. It would be the "evening star" of Mars, visible under the conditions that we see Venus. Earth to the tourist on Mars, however, would not be as bright as Venus is to us; it would be no brighter, at best, than Sirius. Yet Earth would have this advantage over Venus; Earth has an attendant Moon. As seen from Mars, our Moon would have a maximum magnitude of 3.0 It would resemble an average star in brightness and would be clearly visible, with a maximum separation from the Earth of half a degree—the apparent width of the Sun as we see it. From evening to evening, or from dawn to dawn, the changing relationship of Earth and Moon would form a dramatic picture. And, of course, the tourist would be watching "home.")

However, why view a moon and other objects from a planet, when you can view those other objects, plus a planet, from a moon? The view of Earth from the Moon is much more impressive than vice versa and the same can be said, in spades, of the view of Mars from Phobos.

In fact the view of Mars from Phobos is simply tremendous. Mars has a diameter of 4200 miles, only slightly over half that of the Earth, but from Phobos it is seen at a distance of only 3600 miles, surface to surface. Mars is a bloated object in Phobos's sky, 42 degrees of arc from edge to edge, or, to put it another way, if one edge of Mars touched the horizon, the other would be halfway to zenith.

Phobos, in all probability, keeps one side facing Mars at all times, so the red planet would keep its bloated bulk in position, gleaming with a light equal to over 7000 times that of our full Moon. *There's* something for poets to write about and something big enough for lovers' heads to be outlined against.

As an aside, the gravity on Mars is only ⅖ that of the Earth and on Phobos it is virtually nil.

Is there any sight in the Solar System more overpowering than that of Mars as seen from Phobos? Well, to begin with, there is the sight of Jupiter as seen from its nearest satellite.

Jupiter's massive gravitational field will make it a ticklish planet to approach, but no doubt it can be done by working our way slowly down the line of its moons. We can land on one of the outermost satellites (which are only captured planetoids, 15 miles in diameter or so) and build up a base from which a ship can be sent to Callisto.

Callisto is the outermost of the giant Jovian satellites, 1,170,000 miles from Jupiter, but from it, Jupiter already appears larger and brighter than the full Moon appears to us. Within its orbit is first Ganymede, then Europa, then Io. From Io, the innermost of the giant Jovian satellites, Jupiter has waxed in size until it is nearly 400 times as bright as the full Moon and 40 times as wide.

But there is one satellite, a small one (perhaps 150 miles in diameter) that is even closer to Jupiter than Io is. This innermost satellite, variously called Jupiter V, Barnard's satellite, and Amalthea, is only 66,000 miles from Jupiter's surface, and to be that close to Jupiter is something indeed.

In Amalthea's sky, the planet Jupiter would be 46 degrees wide and its area would be rather larger than Mars as seen from Phobos. To be sure, Jupiter is further from the Sun than Mars is and would be less strongly illuminated, so that its globe as seen from Amalthea would be only ⁹⁄₁₀ as bright as that which Mars presents to Phobos. However, Jupiter is far the grander spectacle. Mars is a quiet, ruddy world, with a bare unchanging surface in view. The view of Jupiter, on the other hand, would be that of its turbulent atmosphere, orange, blue, green, and white in belts of frozen free radicals, mottled with storms and crawling with colossal tornadoes. (The brochures would be lyrical indeed for any of the Jovian satellites.)

In addition, Jupiter's four giant satellites would be in Amalthea's skies. Io, the one closest to Amalthea, would appear somewhat larger than our Moon, the remaining three would appear progressively smaller. Each would move through the sky at its own speed, passing behind Jupiter once each revolution and forming a changing pattern that would be a whispered background to Jupiter's colossal shout.

If we compare the Catskills of the sky so far mentioned,

I should guess that the Moon and Mars would be relatively cheap vacation lands; for the masses, so to speak. Phobos, because of its small size, would be expensive and restricted to people with political pull, perhaps. The moons of Jupiter would range from fairly cheap to quite expensive, depending on how close to Jupiter they were (and how much power it took to get there and leave). But certainly it would be Amalthea that would be the true playground of the millionaires.

I can picture a Las Vegas of a sort under a transparent dome on Amalthea. Over it a seething Jupiter hangs ominously. A small pea-sized Sun would cross the sky and pass behind Jupiter every six hours and the moons would come and go in counterpoint. What could be more beautiful?

Well, one thing, of course. Saturn and its rings.

We can eliminate the planets beyond Saturn. The distances are tremendous, the worlds are dim and uninteresting compared with Jupiter. But Saturn has its rings.

To be sure those are a uniquely beautiful sight, but alas, most of Saturn's satellites don't co-operate. Your first thought might be to travel to the satellite nearest Saturn and get a good look at the rings. That satellite would be Mimas, only about 80,000 miles from Saturn's surface and only 35,000 miles from the outermost edge of the rings.

But alas, Mimas revolves in the plane of Saturn's equator and so do the rings. This means that from Mimas, the rings are seen edge-on at all times. Since the rings are very thin (10 miles thick at the most), seeing them edge-on from any reasonable distance means not seeing them at all. The next six satellites beyond Mimas also revolve in the plane of Saturn's equator and all are likewise useless as far as a good view of the rings is concerned. The orbit of the eighth satellite, Japetus, is tilted, but not enough.

That leaves the ninth and outermost satellite, Phoebe. This is really a captured planetoid (200 miles in diameter, perhaps) and does not rotate in the plane of Saturn's equator. Its orbit, in fact, is inclined 30 degrees to the equator so that the rings may be seen, at times, at a more extreme angle than they can possibly be seen from the Earth.

It is too bad that Phoebe is at a distance of 8,000,000 miles from Saturn. From that distance, Saturn seems no larger than the Moon seems to us. The rings would stretch over an extreme width of only a little more than twice the width of the Moon. And yet let's not complain. Even at that distance, Phoebe would offer what most people would be content to admit was the most steadily beautiful naked-eye sight in the Solar System. (And no doubt the tours will be priced accordingly.)

Phoebe revolves about Saturn in eighteen months, which means that every nine months the rings are seen edge-on, while half way between those two edge-on appearances, there would be a maximum view. The wise tourist would time his visits for the maximum view if he could afford it. Those who had to count the dollars would have to take advantage of the lower rates during the times when the rings approach the edge-on view. And, no doubt, the two weeks before and after the edge-on view will be the "slow season" on Phoebe.

That leaves one tourist view which is probably the most frightening and ferocious of all, too frightening and ferocious ever to be really popular, I dare say. I am referring, of course, to a close look at the Sun.

There are two important bodies in the Solar System from which the Sun would seem larger and brighter than from the Earth. These are Venus and Mercury. Venus can be left out of account. Its clouds effectively block off a view of the Sun and, even if it could be seen, it would only be about 1.8 times the size and brightness it appears from Earth.

Mercury does much better. At its aphelion, Mercury's sun is over 4 times as large and bright as ours is, and at perihelion it is just over 10 times as large and bright. However, Mercury would not be an easy place to reach and I have a feeling that tourist accommodations would always be poor.

What I have in mind, though, is an even more extreme case.

There is a planetoid named Icarus, discovered in 1948, which has a rather cometlike orbit. At the far end of the flattened ellipse in which it travels, Icarus retreats from the Sun to an aphelion distance of 184,000,000 miles (which is 30,000,000 miles further out from the Sun than Mars ever gets.)

As it moves toward perihelion, however, Icarus flashes past the orbit of Earth, of Venus, and even of Mercury, and approaches to within nearly 17,000,000 miles of the Sun, whizzing about it in a fast spin and then heading outward again.

When it is at perihelion, the Sun is almost 30 times as

large and as bright as it seems to us on Earth. The surface of Icarus must glow in red heat when it skims the Sun.

For most of its orbital travels, however, Icarus would be far enough from the Sun for ships to land on it without trouble. Suppose the safe interval was used to blast a cavern inside the mile-wide planetoid. A few thousand feet of rock would stave off the heat of the Sun during the close approach (rock is an excellent insulator) and appropriately filtered and protected television receivers could present a view of the Sun that would be unimaginably magnificent.

Undoubtedly, the Icarus Solar Station would be available only as a scientific laboratory and would *not* be open to tourists. However, occasionally, a Congressman or some other VIP might wangle a trip.

And if so, what a story he'll have to tell.

As for myself, after thinking it over carefully, I think I'll stay home. I've been so many places now, without budging from my chair, that my typewriter is beginning to blur and look like a spaceship. Even that much worries my gentle, unadventurous spirit.

But I'll be glad to stand at the spaceport and wave good-by if any of the rest of you want to go.

7 Beyond Pluto

In the last two centuries, the Solar System was drastically enlarged three times; once when Uranus was discovered in 1781; then when Neptune was discovered in 1846; finally when Pluto was discovered in 1930.

Are we all through? Is there no more distant planet to be discovered even yet? We can't know for sure, but at least we can speculate. That much is our fundamental human right.

So— What might a possible Tenth Planet be like? To begin with, how far ought it to be from the Sun? For the answer, we'll go back to the eighteenth century.

Back in 1766, a German astronomer, Johann Daniel Titius, devised a scheme to express simply the distances of the planets from the Sun. He did this by starting with a series of numbers, of which the first was 0, the next 3, and each one following double the one before, thus:

    0, 3, 6, 12, 24, 48, 96, 192, 384, 768. ...

Then he added 4 to each term in the series to get the following:

    4, 7, 10, 16, 28, 52, 100, 196, 388, 772. ...

Now represent Earth's mean distance from the Sun as 10 and calculate the mean distance of every other planet in proportion. What happens? Well, we can make a small table listing Titius's series of numbers and comparing them with the relative mean distances from the Sun of the six planets known in Titius's time. Here's how it would look:

Titius's series	Relative distance	Planet
4	3.9	1) Mercury
7	7.2	2) Venus

10	10.0	3) Earth
16	15.2	4) Mars
28		
52	52.0	5) Jupiter
100	95.4	6) Saturn

When Titius first announced this, no one paid attention, particularly, except for another German astronomer named Johann Bode. Bode wrote about it in 1772, banging the drums hard on its behalf. Since Bode was much more famous than Titius, this relationship of planetary distances has ever since been referred to as Bode's law, while Titius remains in profound obscurity. (This shows you can't always trust to posterity for appreciation either—a thought which should help sadden us further in our moments of depression.)

Even with Bode pushing, the series of numbers was greeted as nothing more than a bit of numerology, worth an absent smile and a that-was-fun-what-shall-we-play-next? But then, in 1781, an amazing thing happened.

A German-born English astronomer, named Friedrich Wilhelm Herschel (he dropped the Friedrich and changed the Wilhelm to William after becoming an Englishman) was engaged that year in a routine sweeping of the skies with one of the telescopes he had built for himself. On March 13, 1781, he came across a peculiar star that seemed to show a visible disc, which actual stars do not do under the greatest magnification available at that time (or now either, for that matter). He returned to it night after night and by March 19, he was certain that it was moving with respect to the stars.

Well, anything with a visible disc and movement against the stars could not be a star, so it had to be a comet. Herschel announced the body as a comet to the Royal Society. But then, as he continued his observations, he couldn't help noting that it wasn't fuzzy like a comet, but had a sharply ending disc like a planet. Moreover, after he had observed it for a few months, he could calculate its orbit, and that turned out to be not strongly elliptical like the orbit of a comet but nearly circular like the orbit of a planet. *And* the orbit lay far outside the orbit of Saturn.

So Herschel announced that he had discovered a new planet. What a sensation! Since the telescope had been invented nearly two centuries before, a number of new objects had been discovered; new stars and several satellites

for both Jupiter and Saturn; but never, never, never in recorded history had a new planet been discovered.

At one bound, Herschel became the most famous astronomer in the world. Within a year he was appointed private
astronomer to George III, and six years after that he married
a wealthy widow. There was even a move, ultimately de-
feated, to name the planet he had discovered "Herschel."
(It is now called Uranus.)

And yet the discovery was accidental and hadn't even
been really new. Uranus is actually visible to the naked eye
as a very dim "star," so it was casually seen any number of
times. Astronomers had seen it through telescopes and on a
number of occasions its position was even reported. As far
back as 1690 John Flamsteed, the first British Astronomer
Royal, prepared a star map in which he carefully included
Uranus—as a star.

In short, any astronomer could have discovered Uranus if
he had looked for it. And he would have had a good hint
as to what kind of a body to look for and how fast he might
expect it to move against the stars, for he would have known
its distance from the Sun in advance. Bode's law would have
told him. The Bode's law figure for the relative distance of
the Seventh Planet (on an Earth-equals-10.0 scale) is 196
and Uranus's actual distance is 191.8.

Obviously, astronomers weren't going to make this mistake
again. Bode's law was suddenly the guide to fame and new
knowledge and they were going to give it all they had. To
begin with, there was that missing planet between Mars and
Jupiter. At least *now* they realized there must be a missing
planet, for Bode's law had number 28 between the orbits of
Mars and Jupiter and no planet was known to exist there.
It had to be searched for.

In 1800, twenty-four German astronomers set up a kind
of community effort to find the planet. They divided the
sky into twenty-four zones and each member was assigned
one zone. But alas for planning, efficiency, and Teutonic
thoroughness. While they were making all possible prepara-
tions, an Italian astronomer, Giuseppe Piazzi, in Palermo,
Sicily, accidentally discovered the planet.

It was named Ceres, after the tutelary goddess of Sicily,
and proved to be a small object only 485 miles in diameter.
It turned out to be only the first of many hundreds of tiny
planets ("planetoids") discovered in the region between Mars
and Jupiter in the years since. Planetoids numbers 2, 3, and
4, by the way, were found by the German team of astron-

omers within a year or two after Piazzi's initial discovery, so teamwork wasn't a dead loss after all. Ceres is far the largest of all the planetoids, however, so let's concentrate on it. Its relative mean distance from the Sun is 27.7; Bode's law, as I said, calls for 28.

No astronomer was in the mood to question Bode's law after that.

In fact, when Uranus's motion in its orbit seemed to be a bit irregular, a couple of astronomers, John Couch Adams of England and Urbain J. J. Leverrier of France, independently decided there must be a planet beyond Uranus with a gravitational pull on Uranus that wasn't being allowed for. In 1845 and 1846 they both calculated where the theoretical Eighth Planet ought to be to account for the deviations in Uranus's motions. They did that by beginning with the assumption that its distance from the Sun would be that which was predicted by Bode's law. A few more assumptions and both pointed to the same general position of the sky. And the Eighth Planet, Neptune, proved to be there, indeed.

The only trouble was that it turned out they had made the wrong basic assumption. Neptune ought to have been at relative distance 388 from the Sun. It wasn't; it was at relative distance 301. It was a little matter of 800,000,000 miles closer to the Sun than it should have been, and with one blow that killed Bode's law deader than a dried herring. It went back to being nothing more than an interesting piece of numerology.

When, in 1931, the Ninth Planet, Pluto, was discovered, no one expected it to be at the Bode's-law distance predicted for the Ninth Planet (the numbers of the planets are selected, by the way, by skipping the planetoids so that Mars is the Fourth and Jupiter the Fifth), and it wasn't.

But now wait.

There are four known bodies lying beyond Uranus and every one of them is odd, in one way or another. The four are Neptune and Pluto, plus Neptune's two known satellites, Triton and Nereid.

The oddness of Neptune is, of course, that it lies so much closer to the Sun than Bode's law would indicate. The oddness of Pluto is more complicated. In the first place it has the most eccentric orbit of any of the major planets. At aphelion it recedes to a distance of 4,567,000,000 miles from the Sun, while at perihelion it approaches to a distance of a mere 2,766,000,000 miles. At perihelion it is actually an average of about 25,000,000 miles closer to the Sun than is Neptune.

Right now, Pluto is approaching perihelion, which it will reach in 1989. For a couple of decades at the end of the twentieth century, Pluto will remain closer to the Sun than Neptune, then it will move out beyond Neptune's orbit, heading toward its aphelion, which it will reach in 2113.

A second odd feature about Pluto is that the plane of its orbit is tilted sharply to the ecliptic (which is the plane of the Earth's orbit). The tilt is 17 degrees, which is much higher than that of any other planet. It is this tilt which keeps Pluto from ever colliding with Neptune. Although their orbits seem to cross in the usual two-dimensional representation of the Solar System, Pluto is many millions of miles higher than Neptune at the point of apparent crossing.

Finally, Pluto is peculiar in its size. It is 3600 miles in diameter, much smaller than the four other outer planets. It is also much denser. In fact, in size and mass, it resembles an inner planet such as Mars or Mercury much more than it does any of the outer planets.

Now let's consider Neptune's satellites. One of them, Nereid, is a small thing, 200 miles in diameter and not discovered until 1949. The odd thing about it is the eccentricity of its orbit. At its nearest approach to Neptune, it comes to within 800,000 miles of the planet, then it goes swooping outward to an eventual distance of 6,000,000 miles at the other end of its orbit. Nereid's orbit is by far the most eccentric orbit in the Solar System. No planet, satellite, or planetoid can compare with it in that respect; only comets equal or exceed that eccentricity.

In contrast to Nereid, Triton is a large satellite, with a diameter in excess of 3000 miles (as compared with the 2160 mile diameter of the Moon) and with a nearly circular orbit. The odd thing about it though is that its orbit is tilted sharply to the plane of Neptune's equator; it is quite near to being perpendicular to that plane, in fact.

Now, there are other satellites in the system with eccentric orbits and tilted orbits. They include the seven outermost satellites (unnamed) of Jupiter; and Phoebe, the ninth and outermost satellite of Saturn. Astronomers agree that these outer satellites of Jupiter and Saturn are probably captured planetoids and not original members of the planetary family. The original members (such as the five inner satellites of Jupiter, including the giant satellites, Ganymede, Io,

Callisto, and Europa the eight inner satellites of Saturn, including the giant satellite Titan) all revolve in nearly circular orbits and in the plane of their planet's equator. So, for that matter, do the five small satellites of Uranus and the two small satellites of Mars. From the manner in which satellite systems are supposed to have originated, these circular, untilted orbits are inevitable.

Well, perhaps Nereid represents a captured planetoid, although it is surprising that a planetoid is to be found so far beyond the planetoid belt, especially one so large (there are not more than four or five planetoids, at most, that are as large as Nereid). And as for Triton, was it captured too? What would an object as large as Triton be doing wandering around in the region of Neptune, getting captured?

Some astronomers have suggested that a catastrophe took place, during some past age, in the neighborhood of Neptune. They suggest that Pluto, which is much more nearly the size of a satellite than the size of an outer planet, was originally indeed a satellite of Neptune. However, it was somehow jarred out of position and took up its present wild and eccentric but independently planetary orbit. The shock of that catastrophe may also have jarred Triton's orbit into a strong tilt.

But what was the catastrophe? That no one says.

The one obvious sign of a possible catastrophe in the Solar System is, of course, the asteroid belt. There is no real evidence that there ever was a single planet there, but certainly it is tempting to believe that one was there once and that it exploded (due to the tidal forces within its crust induced by its next-door neighbor, the giant planet, Jupiter, perhaps). An explosion which produced some 44,000 fragments of rock including Ceres, which is 485 miles in diameter, and three or four others of 100 miles in diameter or more would certainly be a catastrophe.

One catch, however, is that the total mass of all the planetoids between Mars and Jupiter cannot possibly be more than a tenth that of Mars, or more than a fifth that of Mercury. It would still have been far and away the smallest planet in the system. Why should that be? Was it because its neighbor Jupiter gobbled up most of the raw materials for planet formation, leaving our mythical planet a pygmy?

Or suppose that just a fraction of the original planet remained in the space between the orbits of Mars and Jupiter after the explosion? What if the "4½th planet" (we must call it this since Mars is the 4th and Jupiter the 5th) sent a large piece of itself flying far out into space. We can imagine such a piece sailing far out beyond Jupiter, Saturn, Uranus; being caught or seriously deflected by Neptune.

Perhaps the piece was caught by Neptune in an odd orbit and became Triton, while Pluto, as Neptune's original satellite, was knocked out into an independent but whimsical planetary orbit as a result. Or perhaps the piece of the 4½th planet was deflected into the planetary orbit, becoming Pluto, while its gravitational pull tilted Triton's orbit. Or perhaps all three, Pluto, Triton, and Nereid are fragments of the 4½th planet.

The chief bother in all this is how an explosion of the 4½th planet could send so much material far outward, all in one direction. Could it be that this was balanced by the sending of a roughly equal mass inward, toward the Sun?

This brings up the question of our own Moon. Like Triton, the Moon is tilted to the plane of its primary's equator; not by as much, but by a good 18° and its orbit is moderately eccentric as well. Furthermore, the Moon is far too large for us. A planet the size of the Earth has no business with such a huge moon. Of the other inner planets, Mars has two peewee satellites of no account whatever, while Venus and Mercury have none at all.

The Moon is ⅟₈₀ the mass of the Earth and no other satellite in the system even approaches a mass that large in comparison to its primary.

Is it possible then that the inward-speeding fragment of the 4½th planet was captured by the Earth and became the Moon? It sounds, I admit, very unlikely—but speculation is free. Suppose the Moon fragment split up further as it approached Earth and underwent the stresses of our planet's gravitational field. One piece of the fragment might have been slowed sufficiently to allow capture by the Earth, while the other moved at a speed that allowed it to escape from the Solar System altogether.

Or perhaps, to pile improbability upon improbability, this last piece did not escape but was captured by the Sun, so to speak, and became Mercury, which has, next to Pluto, the most eccentric and the most tilted orbit of all the major planets.

If the Moon, Triton, Pluto, and Mercury are all lumped together with the debris of planetoids that are left in the original orbit, you would have a body which would be rather more massive than Mars. This is a respectable planet that would fit the 4½th position nicely.

Of course, I can't imagine what in all this would account for the fact that Neptune's orbit is so much closer to the Sun than it ought to be, but what the devil, we can't have everything. Let's just leave explanations of the fine points to the astronomers and continue to content themselves with the heady delight of ungoverned speculation. We can suppose that all the bodies beyond Uranus form one complex, to be counted as a single planet, in which the average relationship to the Sun remains what it ought to be, but in which the relationship of the individual pieces has been confused by catastrophe.

If we take the mean distance of the whole complex, that turns out to be (thanks to Pluto) 3,666,000,000 miles which, on the Earth-equals-10.0 basis, comes out to be 395.

Now let's make up a new table of the Titius series, like so:

Titius series	Relative distance	Planet
4	3.9	1) Mercury
7	7.2	2) Venus
10	10.0	3) Earth
16	15.2	4) Mars
28	27.7	4½) Ceres
52	52.0	5) Jupiter
100	95.4	6) Saturn
196	191.8	7) Uranus
388	395	8, 9) Neptune-Pluto
772	?	10) Tenth Planet

There you are, then. To answer the question I asked at the beginning of the article, the Tenth Planet should be at position 772 which means it would have a mean distance of 7,200,000,000 miles from the Sun.

How big would it be? Well, if we ignore the interloping Pluto and just consider the other four outer planets, we find a steady decrease in diameter as we move out from Jupiter. The diameters are 86,700 (Jupiter), 71,500 (Saturn), 32,000 (Uranus) and 27,600 (Neptune). Carry that through and let's say the Tenth Planet has a diameter of 10,000 miles, which makes a nice round figure.

With that diameter and at that distance from the Sun (and from us) the Tenth Planet ought to have an apparent magnitude of 13, which would make it rather brighter than the nearer but smaller Pluto. It would show very little disc, but what disc there was would be larger than that of the nearer but smaller Pluto. Well, then, since Pluto has been discovered and the presumably larger and brighter Tenth Planet has not, does that mean the Tenth Planet does not exist?

Not necessarily. Pluto was recognized among a veritable flood of stars of its magnitude or brighter by the fact that it moved among them. So would the Tenth Planet, but at a much slower rate. From Kepler's third law, we can calculate that the period of revolution of the Tenth Planet would be 680 years, nearly three times the length of Pluto's period of revolution, and so the Tenth Planet would move at only one third the rate at which Pluto moves against the stars. It would take a full year for the Tenth Planet to shift its position over the width of the full Moon. This is not the kind of motion that is easily observed by a casual survey of the heavens. Or perhaps it has been seen a number of times and not noticed, as Uranus was.

The thing that strikes me as most unusual about the Tenth Planet is its utter isolation. It is twice as far from Neptune, at Neptune's closest approach to it, as we on Earth are. Most of the time, it is further from Pluto than we are. Once every 2700 years, allowing the most favorable conditions, Pluto would approach within two and a half billion miles of the Tenth Planet (the distance from Earth to Neptune). Nothing else, barring a possible satellite or comet, would ever come within four and a half billion miles of it.

The Sun would have no discernible disc to the naked eye, of course. It would seem completely starlike and no larger in appearance than the planet Mars appears to us at the time of its closest approach. However, although the Sun would be but a point of light, it would still be over sixty times as bright as our full Moon, and a million times brighter than Sirius, the next brightest object in the sky.

If there were any sentient beings native to the Tenth Planet, that alone ought to tell them there was something different about this particular star. Furthermore, if they watched closely, they would see that the Sun constantly, if slowly, shifted position against the other stars.

As to the planets, all the known members of the Solar System, as seen from the Tenth Planet, would seem to hug the Sun. Even Pluto, viewed from so far beyond its own orbit, would never depart more than 40° from the Sun, even when it happened to be at aphelion at the time of maximum elongation. All other planets would remain far closer to the Sun at all times.

As seen from the Tenth Planet, Mercury and Venus would never be more distant from the Sun than the diameter of our full moon. The Earth would recede, at times, to a distance that was at most half again the width of the full moon and Mars would periodically recede to a distance twice the width of the full moon. I feel certain that, even in the absence of an obscuring atmosphere, all four planets would be lost in the brilliance of the point-size Sun and would never be seen from the Tenth Planet without special equipment.

That leaves only the five outer planets, Jupiter, Saturn, Uranus, Neptune, and Pluto. They would be best seen when well to one side of the Sun at which time they would show up (in telescopes) as fat crescents. In that position, Jupiter, Saturn, Uranus, and Neptune would all be at roughly the same distance from the Tenth Planet. Pluto might, under favorable conditions, be rather closer than the rest.

This means that, with the distance factor eliminated, Saturn would be dimmer than Jupiter, since Saturn is smaller and more distant from the Sun, hence less brightly illuminated. By the same reasoning, Uranus would be dimmer than Saturn, Neptune would be dimmer than Uranus, and Pluto dimmer than Neptune.

In fact, Uranus, Neptune, and Pluto, although approaching more closely to the Tenth Planet at inferior conjunction, than do Jupiter and Saturn, would be invisible to the naked eye.

Jupiter and Saturn would be the only planets visible from the Tenth Planet without special equipment and they would be anything but spectacular. At its brightest, Jupiter would have a magnitude of something like 1.5, about that of the star Castor. And it would only be for a year or so, every six years, that it would approach that brightness, and then it would be only 4° from the Sun and probably not too easy to observe. As for Saturn, there would be two-year periods every fifteen years when it might climb to a brightness of 3.5, about that of an average star. That's all.

Undoubtedly, any astronomers stationed on the Tenth Planet would completely ignore the planets. Any other world in the system would give them a better view. But they would watch the stars. The Tenth Planet would offer them the largest parallaxes in the system, because of its mighty orbital sweep. (Of course, they would have to wait 340 years to get the full parallax.) Measurements of stellar distance by parallax, the most reliable of all methods for the purpose, could be extended one hundred times deeper into space than is now possible.

One last point. What ought we to name the Tenth Planet? We've got to stick to classical mythology by long and revered custom. With the Ninth Planet named Pluto, there might be a temptation to name the Tenth after his consort Proserpina, but that temptation must be resisted. Proserpina is the inevitable name for any satellite of Pluto's that may ever be discovered and should be rigidly reserved for that.

However, consider that the Greeks had a ferryman that carried the souls of the dead across into Hades, the abode of Pluto and Proserpina. His name was Charon. There was also a three-headed dog guarding the entrance of Hades, and its name was Cerberus.

My suggestion then is that the Tenth Planet be named Charon and that its first discovered satellite be named Cerberus.

And then any interstellar voyager returning home and approaching the Solar System on the plane of the ecliptic would have to cross the orbit of Charon and Cerberus to reach the orbit of Pluto and Proserpina. What could be more neatly symbolic than that?

8 Steppingstones to the Stars

There's something essentially unsatisfactory to me about the conquest of the Solar System which now seems to be at hand. We know too much about what we'll find, and what we'll find won't be enough.

After all, except for some possible lichenlike objects on Mars, the other worlds of the Solar System are all barren (barring a most unexpected miracle).

Sure, we'll get all sorts of information and knowledge. In the process of reaching these barren worlds, we'll develop valuable alloys, plastics, fuels. We'll work up useful techniques of miniaturization, automation, and computation. I wouldn't minimize any of these advances.

But—there will be no Martian princesses, no tentacled menaces, no superhumanly intelligent energy beings, no dreadful monsters to bring back to zoos. In short, there won't be any romance!

For the proper results and rewards of space travel, we must reach the stars. We must find the Earth-type planets that possibly circle them, carrying upon them their full complement (we hope) of friend and foe, of superman and monster.

Only how do we get to the stars? The Moon may be on our doorstep and Mars may be just across the threshold, but the stars are way to helengone out of sight.

The Moon is 222,000 miles away at its nearest and Mars is 35,000,000 miles away at its nearest. Even Pluto, the most distant of the known planets, is never further than 4,650,000,000 miles from us. On the other hand, the Alpha Centauri system, which includes the nearest stars to us, is 25,000,000,000,000 miles away.

In other words, when we've labored our way to the farthest edge of the Solar System and stand on Pluto, we have covered a distance which is, at best, less than ⅟₅₀₀₀ of the distance that must be covered if even the nearest star is to be reached.

It would be so nice if there were steppingstones to the stars; if there were bodies between Pluto and the stars that would at least give us a breathing spell, a place to stop and rest on the long trip to the nearest stars.

And having said that, I can smile cheerfully and say that there is good reason to believe that such steppingstones do exist. I don't mean dark stars which may or may not exist between us and Alpha Centauri; and I don't mean trans-Plutonian planets, which may or may not exist.

I am referring, rather, to a shell of planetoids which surrounds the Sun, far beyond Pluto's orbit, with a dark halo; a shell of planetoids that dwarfs the known Solar System and which, in all probability, actually exists.

To tell the story of these planetoids, I shall, as is my wont, begin at the beginning. In this case, the beginning involves the comets.

From time immemorial, comets have been considered portents of disaster, and with what seemed good reason.

After all, the heavens are, for the most part, a scene of quiet changelessness or, at most, of majestically periodic change. The sun rises and sets, the moon runs through its phases, the "fixed" stars maintain their positions exactly from generation to generation, and the planets wander among them in complicated but predictable paths.

All is well. All is peaceful.

Then, hurrying into view, apparently from nowhere, comes a comet. It is like nothing else in the heavens. A fuzzy patch of light, the "coma," surrounds a bright starlike nucleus, and extending from the coma is an arched tail that can stretch halfway across the heavens. Having come from nowhere, the comet finally vanishes into nowhere. There seemed no way of predicting either its coming or going and all one could say was that it had disturbed the peace and serenity of the skies.

This was in itself disturbing enough. Add to that the strangeness of its shape. It resembled a distraught woman, tearing across the sky in a hysterial frenzy, her unbound hair streaming behind her in the wind. The very word "comet" comes from the Greek *kometes* meaning "long-haired."

Naturally, any sensible man could only suppose that such a sudden and frightening apparition was sent by some god

to warn humanity of disaster. And furthermore, since life
and humanity are such that disaster strikes every year without fail, this theory seems to be borne out unmistakably.
After a comet, disaster invariably follows. Within a year
of the comet's appearance, there is sure to be a war, plague,
or famine somewhere, or some famous man dies, or some
heretic appears or something.

The last halfway spectacular comet showed up in 1910,
and it succeeded in frightening many people into believing
the end of the world would surely come. (It also, as any
fool can plainly see, foretold the death of Mark Twain, the
sinking of the *Titanic,* the coming of World War I, and a
whole slew of catastrophes.)

However, portent or not, what is the nature of a comet?
Aristotle, and the ancient and medieval thinkers who followed him, believed the heavens were perfect and unchange-
able. Since comets came and went, having a beginning and
an ending (which stars and planets did not) they were imperfect and changeable and, therefore, could not be part of
the heavens. They were instead atmospheric phenomena; ex-
halations of bad air and therefore part of our own corrupt
and miserable Earth.

This notion was not destroyed until 1577. The Danish
astronomer, Tycho Brahe, measured the parallax of a bright
comet that appeared that year, plotting its position as seen
against the stars from his own observatory in Denmark and
from another observatory in Prague. The parallax proved too
small to measure. This is not surprising, considering the relative shortness of the base line (about 500 miles) and the fact
that this was before the days of the telescope. However,
even so, if the comet had been within 600,000 miles of the
Earth, its parallax could have been perceptible. Tycho's conclusion, then, was that the comet had to be *at least* three
times as far from the Earth as the Moon was. That made that
comet, at any rate, part of the heavens; and Aristotle was
wrong.

Even as part of the heavens rather than of the Earth,
comets remained troublesome. They didn't fit into any system.
When Copernicus put the Sun at the center of the Solar
System and Kepler made planetary orbits into ellipses, the
design of the planets began to fall neatly into place—except
for the comets. They still came from nowhere, vanished into
nowhere, and represented an irritating lawlessness in the
kingdom of the Sun.

Then came Newton and his law of gravitation that so neatly explained the planetary movements. Could it also explain cometary movements? That would indeed be an acid test.

In the year 1704 Edmund Halley, a good friend of Newton, began to work out the orbits of various comets over the regions for which observational records existed, in order to see if their motions could be made to fit the requirements of gravitational mathematics. The records of twenty-four different comets were studied.

The one with the best available data was the comet of 1682, which Halley had himself observed. Working out its orbit, he noticed that it passed through the same regions of the sky as had the comet of 1607, seventy-five years before, and the comet of 1531, seventy-six years before that. Checking back, he found records of another comet in 1456, seventy-five years further back still.

Could it be that the same comet was coming back at intervals of seventy-five years or so, after passing over an elliptical orbit so eccentric that its far end reached out way beyond the orbit of Saturn, then the furthest planet known?

Halley felt certain that just this was indeed so, and consequently predicted that the comet of 1682 would return once again in 1758.

It is one of the frustrations of scientific history that Halley knew he was not likely to live to see his prediction verified or exploded. He would have had to live to be one hundred and two for that, and he didn't. He made a valiant try, reaching the age of eighty-five, but that wasn't good enough.

On Christmas night 1758 a comet was sighted and through early 1759, it rode high in the sky. The comet had indeed returned and it has been called Halley's comet ever since. (It was Halley's comet that was in the sky in 1910.)

This created a sensation. Comets, or at least one comet, had been reduced to a commonplace, law-abiding member of the Solar System. Since then, many others have been supplied with definite orbits. And now, at last, there is no logical reason for considering comets divinely sent portents of disaster—which, however, will not prevent people preparing for the end of the world at the next appearance of a large comet, you may be sure.

Granted that comets are ordinary members of the Solar System, subject to the same laws of motion as are the sedate planets, what are they? Well, they aren't much.

Comets have frequently approached one or another of the various planets and have had their orbits altered, sometimes drastically, as a result of the gravitational attraction of the planet. (Such perturbations make it rather difficult to pinpoint the time of a comet's return.) The planet, for its part, has never in any way showed any measurable effect due to the comet's gravitational attraction. The comet of 1779 actually passed through Jupiter's satellite system without affecting the satellites in any way.

The obvious conclusion is that for all their gigantic volumes, and some comets are actually more voluminous than the Sun, comets have very small masses. The mass of even a large comet can be no larger than that of a middle-sized planetoid.

If this is so, the density of a comet must be extremely low, far lower than the density of the Earth's atmosphere. This is demonstrated by the fact that stars can be seen through the tail of a comet with no perceptible diminution in brightness. The Earth passed through the tail of Halley's comet in 1910 and there was no discernible effect. In fact, Halley's comet passed between the Earth and the Sun and the whole thing disappeared. The Sun shone through it as though it were a vacuum.

Professor Fred Whipple of Harvard originated, some years ago, a now widely accepted theory of the composition of comets that accounts for all this. Comets, he supposes, are made up largely of "ices," that is, of low-melting solids such as water, methane, carbon dioxide, ammonia, and so on. When far from the Sun, these substances are indeed solid and the comet is a small, solid body. As it approaches the Sun, however, some of the ices evaporate and the dust and gas that form are forced away from the Sun by the radiation pressure of sunlight.

Sure enough (as was first observed in 1531) a comet's tail always points generally away from the Sun. It streams out behind the comet as the comet approaches the Sun, but it precedes the comet as it moves away from the Sun. Moreover, the closer to the Sun, the larger the tail.

Not as much atmosphere is formed, driven away by radiation pressure and lost, as you might think. The ices themselves are poor conductors of heat and comets remain in the vicinity of the Sun only a comparatively short space of time. They retreat with most of their substance intact.

Nevertheless, at each return a comet does lose some of its

substance. Whatever passes into the tail vanishes into space and never returns. A few dozen passes at the Sun would probably suffice to finish a comet. Even a comet that returns only at intervals of a century or so can't be expected to last more than several thousand years at best. Therefore we ought, within historical times, to see comets shrivel and die.

And we do. Halley's comet at its return in 1910 was disappointingly dim, when compared with previous descriptions. It will probably be even more disappointing at its next scheduled appearance in 1986. It is dying.

And some comets have actually died as men watched. The best known example is that of Biela's comet, first discovered in 1772 by the German astronomer, Wilhelm von Biela. It had a period of about 6.6 years and was observed on a number of its returns. In 1846, it was found to have split in two, the halves traveling side by side. In 1852 the two parts had separated further. And Biela's comet was never seen again. It had died.

But that's not the end of the story. Traveling in the orbit of the comet are a group of meteorites. We know, because in 1872 Biela's comet would have passed fairly close to the Earth if there had still been a Biela's comet. There wasn't, but that year we were treated with a meteor shower radiating out of the spot where the comet would have been located.

Apparently, embedded in the ices of the comet are a vast number of pebbles and pinpoints or less of metal and silicates. When the binding ices are gone, the contents fall apart. The small meteors and micrometeors that fill space now may thus be the ghosts of comets long dead.

Obviously, if comets have such short lifetimes and are still as numerous as they are (several new ones are discovered every year) even though the Solar System has been in existence for five billion years, a continual supply must be entering the system. But where are they coming from, then?

The easiest answer is that they come from interstellar space. They may be wanderers among the stars. Some may occasionally enter the gravitational field of the Sun, flash around it and go forever. Some enter, are captured by planets, and become periodic comets, doomed to a quick death.

There are a couple of arguments against this possibility. First, to have interstellar migrants blundering into our Solar System at the rate they do would require the filling of interstellar space with a most unlikely number of comets. Besides, more would enter the system from the direction toward which the Sun is traveling than from the other. That, however, is not so. Comets come from all directions equally.

Secondly, if comets entered the system randomly from outer space, a number should come and go in distinctly hyperbolic orbits (like a hairpin opened wide). No comet with a *distinctly* hyperbolic orbit has ever been observed.

In view of this, a more logical possibility is that the source of the comets is a local reservoir bound to the Sun. It was suggested some years ago that this local reservoir exists in the form of a shell of ice planetoids, located from one to two light-years from the Sun in every direction.

It is easy to see how this shell may have come into existence. If the Solar System began as a vast turbulent cloud of dust and gas some light-years in diameter, then as it swirled and contracted, the planets and present-day Sun would be formed. At the outskirts of the original cloud, however, the density would have been too low for planetary formation and, instead, there would be numerous local concentrations. Since the temperature has remained near the absolute zero throughout billions of years in that far-flung region, the ices which composed much of the original cloud, would be retained even by the tiny gravity of the planetoids. (Nearer the Sun, the higher temperature has caused even as large a body as the Earth to lose much of its supply of ices.)

There is an estimate to the effect that this shell of "cometary planetoids" contains 100,000,000,000 individuals with a mass, all told, up to $1/100$ or even possibly $1/10$ that of the Earth. The average cometary planetoid would then have a mass of 600,000,000 to 6,000,000,000 tons. If we assumed the density of such a planetoid to be equal to that of ice, the average diameter would run, roughly, close to a mile.

You might think that a shell of a hundred billion planetoids ought somehow to make its presence known to observers on Earth. However, consider that the shell of space enclosing the Sun at a distance between one and two light-years, has a volume of thirty cubic light-years. This is immense! If the 100 billion cometary planetoids were evenly distributed through that volume, the average distance separating them would be about $1\frac{1}{4}$ billion miles, which is nearly the distance between ourselves and Uranus.

Naturally, a volume of space containing a mile-wide hunk of ice every billion miles or so is not going to make any impression at all at a distance of a light-year or more. The cometary planetoids will reveal themselves neither by luminosity nor by blocking the light of the stars.

Imagine a cometary planetoid somewhere in the middle of the shell, say 1½ light-years from the Sun. The Sun, from that distance, would seem merely a star, though still the brightest star in the sky, with a magnitude of −2. The planetoid would still be within the gravitational influence of the Sun (no other star would be as close) but that influence would be weak.

A cometary planetoid, 1½ light-years from the Sun, and traveling in a circular orbit about the Sun, would be whipped along under the feeble gravitational lash at a speed of only a little over 3 miles a minute. This may sound fast to the automobile driver, but the Earth moves along its orbit at a rate of 1100 miles a minute and even far-off Pluto never moves at a rate of less than 150 miles a minute.

At its slow rate of movement, it takes the average cometary planetoid 30,000,000 years to complete a revolution about the Sun. In all the existence of the Solar System, those far-distant planetoids have not, on the average, yet had time to revolve about the Sun 200 times.

But if the cometary planetoids are circling quietly way out there, why do they not continue to circle there forever? What sends them down toward the Sun? The only possible answer seems to involve the interfering gravitational influence of the nearer stars. After all, the gravitational pull of Alpha Centauri on those cometary planetoids which happen to be directly between that star and the Sun, is 10 per cent that of the Sun and that is not negligible. (Remember, Alpha Centauri is scarcely farther from some of those planetoids than the Sun is.) A few other stars exert gravitational attractions for those planetoids nearest them to an amount of over 1 per cent that of the Sun.

Now then, if these stellar attractions catch a particular planetoid in such a way as to slow its orbital velocity, it must fall in toward the Sun, its circular orbit becoming elliptical. If the orbital velocity is slowed sufficiently, it must fall in toward the Sun so sharply as to enter the Solar System proper. It would gather speed as it did so, whip around the Sun, and climb back to the point where the perturbation had taken place, then whip down again, climb back, whip down again, and so on. If it came close enough to the Sun, it would develop a gigantic tail and coma of evaporating ices and would become visible to watchers on the Earth.

If only the Sun and the comet existed, this new, highly elliptical orbit would be permanent (barring additional stellar perturbation). A comet traveling in such an orbit would have a much shorter year than it did when it was in its shell, but its year would still be long by Earthly standards—about 10,-000,000 years or so.

As far as man is concerned, such "long-period comets" would be one-shots. Any comet of this type appearing during historical times would not have been viewed by man on its previous visit, for he did not then exist. Moreover there is a distressingly good chance that man may no longer exist to see the next visit.

Of course, once a comet enters the Solar System proper, there is always the chance that it will come close enough to some planet to have its orbit affected. In some cases, its velocity will be speeded so that its orbit will become slightly hyperbolic and it may then leave the Solar System for good. In other cases, its velocity will be slowed and it will no longer gain the kinetic energy required to send it back to the cometary shell. It will often only recede no further than the neighborhood of the planetary perturbation, so that it will, in effect, have been captured by the planet.

All the outer planets have "families" of comets, that of Jupiter, very naturally, being the largest. Perhaps the most remarkable of the Jupiter family is Encke's comet, the orbit of which was worked out in 1818 by the German astronomer, Johann Franz Encke, after it had been discovered by the French astronomer, Jean Louis Pons.

Encke's comet has the shortest period of any known comet —3.3 years. It never recedes further from the Sun than about 400,000,000 miles which means that even at its most distant, it is never as far from the Sun as Jupiter is. It approaches fairly close to Mercury's orbit at its perihelion and its perturbation by Mercury has been used to calculate the mass of that small planet.

As you might expect, Encke's comet is dim and unspectacular, and it never develops a tail. It has been near the Sun far too many times to be anything else. Most of its ices are undoubtedly gone and it must now consist largely of a fairly compact silicate residue, thinly interlarded, perhaps, with the remnant of the original ices.

Naturally, the cometary shell is being depleted by these stellar perturbations. Any cometary planetoid slowed and sent down into the Solar System proper is condemned to death. In addition, other cometary planetoids are speeded by stellar perturbations and may be forced into a hyperbolic orbit that drives them away from the Sun altogether.

On the other hand, no cometary planetoids are being added to the shell as far as we know, so that the number continually declines.

However, this need not be a source of worry. It has been estimated that perhaps three new comets are sent hurling into the Solar System proper each year. We can suppose that three more are, on the average, speeded into hyperbolic loss in each year. At any rate, in the entire five-billion-year history of the Solar System, 30,000,000,000 cometary planetoids have been lost or destroyed. That amounts to only 30 per cent of the estimated number that still remains.

Despite the cometary death rate, then, our comets will be with us in their usual numbers for billions of years more.

It is these cometary planetoids, to get back to the remarks I made at the very beginning of the article, which may represent the steppingstones to the stars.

If we could ever reach Pluto, it might not be too great a hop to reach one of the closer cometary planetoids; one that had been slowed into a relatively skimming approach toward the outskirts of the Solar System proper. Certainly not as great an effort would be required to reach such a planetoid as would be required to reach Alpha Centauri in one jump.

If a base could be set up on such a mile-wide hunk of ices, perhaps we could continue to press outward into space from planetoid to planetoid in an island-hopping fashion, to the outermost fringes of the shell.

Nor would the two-light-year mark necessarily end such island-hopping possibilities. After all, there is no reason to believe that Alpha Centauri doesn't have a halo of cometary planetoids of its own. Why shouldn't it have one? (Though perhaps a more complicated one, since Alpha Centauri is really three stars.)

If it has one, then Alpha Centauri and the Sun are close enough so that the outermost fringes of the halo of one ought to be rather close to the outermost fringes of the halo of the other.

Perhaps, then, we could island-hop over the ice all the way. Perhaps at no point will an uninterrupted trip of more than a few billion miles be required and perhaps we can reach the nearest star, at least, in the way a mountain climber scales a high peak—by establishing a series of inter-mediate bases on the way.

I cannot honestly say that this makes a trip to the stars actually look inviting, but if we've *got* to go, surely it is easiest to go a step at a time.

9 *The Planet of the Double Sun*

The title sounds as though this were going to be a rather old-fashioned science-fiction story, doesn't it?

Yet although the title may sound old-fashioned, the situation need not be. One of the most glamorous settings that can be imagined is that of more than one sun in the sky.

The author of a story describing such a setting need not (and usually does not) worry about the astronomic verities of the situation. The suns are usually described as looking like suns and both (or all) are made to move independently in the sky. The author will usually throw in local color by saying that one sun was just rising, while the other had just passed zenith. He may make matters more colorful (figuratively *and* literally) by having one sun, for example, red and the other blue. Then he can talk of double shadows and their various configurations and color combinations.

A little of this is enough to make us sigh at our misfortune in having only one sun in the sky; and a pretty colorless one at that. Oh, the missing glories!

What *would* it be like to have more than one sun in the sky? There are, of course, a wide variety of types of multiple stars; some are made up of two components and some of more than two. In some multiple stars, the components are near together; in others far apart. The components may be similar or not similar; one may be a red giant or one may be a white dwarf.

But let's not make up any systems or look for something exotic or foreign. The fact of the matter is that we have an example in our back yard. The nearest star to us in space, a star so close we can almost reach out and touch it, a next-door neighbor no more than 25,000,000,000,000 miles away, good old Alpha Centauri, is a multiple star.

Suppose we were on a planet in the Alpha Centauri system. What would it be like?

To begin with, what is Alpha Centauri like?

In the first place, Alpha Centauri is a star in the Southern Celestial Hemisphere. It is never visible in the sky north of about 30 degrees north latitude. The chances are you've never seen it; I know I never have. Moreover, the ancient Greeks never saw it.

The chief observatories of the medieval Arabs, in Cordova, Baghdad, and Damascus, were all north of the 30-degree line. Presumably ordinary Arabs in the Arabian and Sahara deserts must occasionally have seen a bright star very near the southern horizon, but this, apparently, did not penetrate to the egghead level.

The test of the matter is that Alpha Centauri, although the third brightest star in the sky, has no name of its own, neither Greek nor Arabic. (The name Alpha Centauri is official "astronomese.")

Of course, once Europeans started adventuring down the coast of Africa in the late 1400s, the bright star must have been observed at once. Eventually, astronomers got around to making star maps of those parts of the Southern Celestial Hemisphere invisible from Europe. (The first was Edmund Halley of Halley's comet fame who, in 1676, at the age of twenty, traveled to St. Helena of future Napoleonic fame to map northern stars.) Astronomers divided the southern heavens into constellations to complete the scheme already begun in those parts of the heavens which the ancients had been able to observe.

They named the constellations in Latin, naturally, and included mythological creatures as a further match to what already existed in the sky (just as planets discovered in modern times received mythological names matching the older ones). One of the prominent southern constellations was named the Centaur. In Latin, this is *Centaurus* and the genitive ("of the Centaur") is *Centauri*.

Centaurus contains two first magnitude stars. The brightest was named Alpha Centauri and the other Beta Centauri. The words "alpha" and "beta" are not only the first two letters of the Greek alphabet but were also used by the Greeks to represent the numbers "one" and "two," a habit never broken by scientists. The names of the stars, freely translated, therefore mean "star number one of the Centaur" and "star number two of the Centaur" respectively.

The magnitude of Alpha Centauri is 0.06 which makes it,

as said, the third brightest star of the sky. The only stars brighter are Canopus (−0.86) and, of course, Sirius (−1.58).

(The lower the magnitude, the brighter the star, in a logarithmic ratio. A difference in magnitude of one unit means a difference in brightness of 2.512 times. A difference in magnitude of two units means a difference in brightness of 2.512 × 2.512 or about 6.31 times and so on.)

About 1650 telescopes became good enough to detect the fact that some stars, which looked like single points of light to the naked eye, were actually two closely spaced points of light. In 1685 Jesuit missionaries in Africa, taking time out for astronomical observations, first noticed that Alpha Centauri is an example of such a double star. The brighter component is Alpha Centauri A, the other Alpha Centauri B.

The magnitude of Alpha Centauri A by itself is 0.3 and that of Alpha Centauri B is 1.7. The 1.4 difference in magnitude means that Alpha Centauri A is 3.6 times as bright as Alpha Centauri B. To translate the brightness into absolute terms; that is, to compare either component with our Sun, it is necessary to know the distance of Alpha Centauri.

This distance could be measured by noting slight shifts in the star's position, mirroring the change in Earth's position as it revolved about the Sun. This tiny yearly motion of a star, resulting from Earth's motion, is called stellar parallax and grows smaller as the distance of a star increases. A very distant star has virtually no parallax at all, so it can be treated as a motionless reference point against which the parallax of a nearby star could be measured. (Without some reference point, parallax is meaningless.)

However, astronomers had for centuries been trying to detect smaller parallaxes without success, although they had succeeded first with the parallax of the Moon, then of the Sun and the planets. Apparently, even the nearest stars had parallaxes so small as to make them difficult to measure.

Another trouble was that without knowing the parallaxes, one couldn't tell which star was near and which far. How, then, know which star to measure and which to use as a motionless reference point?

Astronomers made the general assumption that, all things being equal, a bright star is closer to Earth than is a dim star. Also, a star with high proper motion (a shift in position due to the star's own motion through space; a shift which is continuous—always in one direction and not cyclic, or back and forth, as parallactic shifts would be) was assumed nearer the Earth than one with a low proper motion. These assumptions would not necessarily hold true in every case, for a bright star might be more distant than a faint one, but be enough brighter, intrinsically, to make up for that. Again, a near star might have a very rapid apparent motion, but one which was in our line of sight so that it wouldn't show up. Nevertheless, these assumptions at least gave astronomers a lead.

By the 1830s the time was ripe for a concerted attack on the problem. Three astronomers of three different nations tackled three different stars. Thomas Henderson (British) observed Alpha Centauri and Friedrich Wilhelm Struve (German-born Russian) worked on Vega, the fourth brightest star in the sky. Both stars were not only bright but had pretty snappy proper motions. Friedrich Wilhelm Bessel (German) applied his efforts to 61 Cygni. This was a dim star but it had an unusually high proper motion. In each case, the star's position over at least a year was compared with that of a dim and presumably very far-off neighbor star.

Sure enough, each of the three stars being investigated shifted position slightly compared to its presumably distant neighbor. And so it happened (as it often does in science) that after centuries of failure, there were several almost simultaneous successes.

Bessel got in first, in 1839, and he gets the credit of being the first to measure the distance of a star. It turned out that 61 Cygni is 11 light-years distant. Henderson, later in 1839, reported Alpha Centauri to be a little over 4 light-years distant and Struve, in 1840, placed Vega at about 27 light-years distant.

No star has been found closer than those of the Alpha Centauri system.

Knowing the distance of Alpha Centauri, it is easy to calculate that Alpha Centauri A (the brighter of the pair) is almost exactly as bright as our Sun. Since its spectrum showed it to be at the same surface temperature, it is our Sun's twin—same diameter, same mass, same brightness, same everything, apparently.

As for Alpha Centauri B, if it were the same temperature as Alpha Centauri A, then it would be just as luminous per unit area. To be only $\frac{1}{3.6}$ as luminous as its companion, it must have $\frac{1}{3.6}$ its area. The diameters of the two stars would be as the square roots of the respective areas and (assuming the two stars to be equally dense) the masses would be as the cube of the square roots of the respective areas.

It would then turn out that Alpha Centauri A would have

a diameter 1.9 times that of Alpha Centauri B and a mass about 7 times that of Alpha Centauri B. (Actually, Alpha Centauri B is a trifle cooler than Alpha Centauri A, so that the comparison is not exactly as I've given it, but for the purposes of this essay, we needn't worry about the refinements.)

The two stars rotate in elliptical orbits about a common center of gravity. The period of rotation is about 80 years. When the stars are closest, they are about a billion miles apart. When they are furthest they are 3.3 billion miles apart.

Now, then, suppose we try to duplicate (in imagination) the Alpha Centauri system here in our own Solar System. Since Alpha Centauri A is the twin of our Sun in every respect, let's suppose our Sun is Alpha Centauri A, but let's keep on referring to it, for convenience's sake, as the Sun.

Let's imagine Alpha Centauri B (which we will call simply Sun B) in orbit about the Sun. We can avoid unnecessary complications by making it exactly one half the diameter of the Sun and equally dense so that it is one eighth the Sun's mass. This may not be exactly the situation with respect to Alpha Centauri B, but it is a reasonably close approximation.

Furthermore, let's suppose Sun B is traveling in a nearly circular orbit, in the same plane as the planets generally, and at the average distance of Alpha Centauri B from Alpha Centauri A (again a change in detail but not in essence). This would place it in orbit about 2,000,000,000 miles from the Sun. This is almost as though we have taken the planet Uranus of our Solar System and replaced it with Alpha Centauri B.

All this would make Earth part of a multiple-star system very closely resembling that of Alpha Centauri. Now what would the heavens be like?

In some ways, our Solar System would be changed. Uranus, Neptune, and Pluto, as we know them, would be out. Their orbits would be tangled with Sun B. However, these planets were unknown in the pre-telescopic era, so we can do without them as far as naked-eye observation is concerned.

But even Saturn, the outermost of the planets known to the ancients, would be nearer to the Sun than to Sun B in the position I have placed the latter. With the Sun, on top of that, having a gravitational field eight times as intense as that of Sun B, it should hang on to Saturn and the still closer planets with no trouble. (There might be interesting minor effects on the planetary orbits but I'm not astronomer enough, alas, to be able to calculate them.)

Sun B would behave like a new and very large "planet" of the Sun. The Sun and Sun B would revolve about a center of gravity which would be located in the asteroid belt. The motion of the Sun about this point once every eighty years would, however, not be detectable in pre-telescopic days, because the Sun would carry all the planets, including Earth, with it. Neither the Sun's distance nor Sun B's distance from Earth would be affected by that motion.

(After the invention of the telescope, the Sun's swing—with us in tow—would become noticeable through its reflection in the parallactic displacement of the nearer stars.)

But what would Sun B look like in our heavens?

Well, it would *not* look like a Sun. It would be a point of light like the other planets. A diameter of 430,000 miles at a distance of 2,000,000,000 miles would subtend an angle of about 45 seconds of arc. Sun B would appear to the naked eye to be just about the apparent size of the smaller but closer Jupiter.

To a naked-eye observer (such as the Greeks or Babylonians) Sun B would be one more point of light moving slowly against the stars. It would be moving more slowly than the others, making a complete circuit of the sky in about 80 years, as compared with 29½ years for Saturn and 12 for Jupiter. From this, the Greeks would—rightly—conclude that Sun B was further from Earth than was any other planet.

Of course, one thing would make Sun B very unusual and quite different from the other planets. It would be very bright. It would have an apparent magnitude of about −18. It would be only ⅟₃₀₀₀ as bright as the Sun, to be sure, but it would still be 150 times as bright as the full Moon. With Sun B in the night sky, Earth would be well-illuminated.

Another thing might be unusual about Sun B; not as a matter of inevitability, as with its brilliance, but as a matter of reasonable probability at least.

As a "planet" of the Solar System, why should it not have satellites, as the other planets have? (Of course, its satellites would be revolving about a Sun and would really be planets, but let's not worry about having a consistent terminology.)

To be sure, since Sun B is much larger than the other planets, it could be expected to have a satellite much larger and more distant from itself than is true for any other planet.

It might, for instance, have a satellite the size of Uranus. (Why not? Uranus would be much smaller compared to Sun B, than Jupiter is compared to the Sun. If the Sun can have Jupiter in tow, then it is perfectly reasonable to allow Sun B to have a planet the size of Uranus.)

Uranus could be circling Sun B at a distance of 100,000,-000 miles. (Again, why not? Jupiter, which is considerably smaller than Sun B and considerably closer to the competing gravity of the Sun nevertheless manages to hold on to satellites at a distance of 15,000,000 miles from itself. If Jupiter can manage that, Sun B can manage 100,000,000.)

If Uranus moved about Sun B in the plane of Earth's orbit, it would move first to one side of Sun B, then back and to the other side, then back and to the first side, and so on, indefinitely. Its maximum separation from Sun B would be about 3 degrees of arc. This is about 6 times the apparent diameter of the Sun or the Moon and such a separation could be easily seen with the naked eye.

But would Uranus itself be visible at that distance from us?

Well, right at the moment, *without* Sun B, Uranus *is* visible. It is 1,800,000,000 miles from the Sun (nearly as far as I have, in imagination, put Sun B) and it has a magnitude of 5.7 which makes it just visible as a very faint star.

But if Uranus were rotating about Sun B, it would be lit up not merely by the dim light of the distant Sun (the reflection of which is all we see Uranus by, actually) but also by the stronger rays of the much nearer Sun B.

The average magnitude of Uranus under these conditions would be 1.7 It wouldn't be as bright as the other planets, but it would be brighter than the North Star, for instance. The glare of the nearby Sun B might make Uranus harder to observe than the North Star, but it should still be clearly visible. (Sun B might have more than one satellite, too, but let's not complicate the picture. One satellite will do.)

The Greeks would thus be treated to the spectacle not only of an unusually and exceptionally brilliant point of light but also to another point of light (much dimmer) that oscillated back and forth as though caught in the grip of the brighter point.

Both factors, brilliance and a visible satellite, would be completely unique. I have a theory that this would have made an interesting difference in Greek thinking, both on the mythological and the scientific level.

Mythology first (since Greek mythology is older than Greek science) and that involves the "synodic period" of a planet. This is the interval between successive meetings of a planet and the Sun, in our sky. Jupiter and the Sun meet every 399 days; Saturn and the Sun every 378 days. Sun B and the Sun would meet in Earth's sky every 369 days. (This is just a measure of how frequently Earth in its revolution manages to get on the other side of the Sun from the planet in question.)

As the planet approaches the Sun it spends less and less time in the night sky and more and more time in the day sky. For ordinary planets this means it becomes less and less visible to the naked eye because it is lost in the Sun's glare during the day. Even the Moon looks washed out by day.

But Sun B would be different. Considering that it is 150 times as bright as the full Moon, it would be a clearly visible point of light even by day. Allowing the use of smoked glasses, it could be followed right up to the Sun.

Now the Greeks had a myth about how mankind learned the use of fire. At the time of creation, man was naked, shivering, and miserable; one of the weakest and most poorly endowed of the animal creation. The demigod Prometheus had pity on the new creature and stole fire from the Sun to give to mankind. With fire, man conquered night and winter and marauding beasts. He learned to smelt metals and developed civilization.

But the anger of Zeus was kindled at this interference. Prometheus was taken to the very ends of the earth (which, to the Greeks, were the Caucasus Mountains) and there chained to a rock. A vulture was sent there to tear at his liver every day, but it left him at night in order that his liver might miraculously grow back and be ready for the next day's torture.

There now. Doesn't all this fit in perfectly with the apparent behavior of Sun B? Every year Sun B commits the crime of Prometheus. It can be seen in the daytime approaching the Sun, the only planet that can be seen to do this. It can only be planning to steal light from the Sun, and it obviously succeeds. After all, isn't that why it is so much brighter than all the other planets, why it is so much brighter even than the Moon?

Moreover, it brings this light to mankind, for when it is in the night sky, it illuminates the landscape into a dim kind of day.

But the planet is punished. It is cast out to the edge of

the universe, further away than any other planet. There is even a vulture tearing at it, in the shape of its clearly visible satellite. While the planet was busy stealing fire from the Sun, no satellite was visible (because it was drowned out by the Sun's glare, of course). Once the planet was hurled to the edge of the universe, though, and became visible in the night sky, its satellite appeared. The satellite swoops toward the bright planet, tearing at it, then moves away to allow it to recover, then swoops in again, and so on in an eternal rhythm.

With all this in mind, isn't it just about inevitable that if Sun B were in our sky, it would be named Prometheus? Or that the satellite would now have the Latin name Vulturius.

Now I'm far too sober-minded and prosaic myself to think outlandish thoughts (as all of you know), but I wouldn't be surprised if some people reading this might not think the parallel is far too close to be accidental. Could it be that such a heavenly situation actually existed and suggested the myth in the first place?

Could it be that the human race originated on a planet circling Alpha Centauri A? Could they have migrated to Earth about fifty thousand years ago, wiping out the primitive Neanderthals they found here and established a race of "true men"? Could some disaster have destroyed their culture and forced them to build up a new one?

Is the Prometheus myth a dim memory of the distant past, when Alpha Centauri B lit up the skies? Was the Alpha Centauri system the original of the Atlantis myth?

No, I don't think so, but anyone who wants to use it in a science-fiction story is welcome to it. And anyone who wants to start a religious cult based on this notion probably can't be stopped but please—don't send me any of the literature—and *don't* say you read it here first.

And what effect would Sun B (or "Prometheus") have had on Greek science?

Well, in the real world, there was a time when matters hung in the balance. The popular Greek theory of the universe, as developed by 300 B.C., put the Earth at the center and let everything in the heavens revolve about it. The weight of Aristotle's philosophy was on the side of this theory.

About 280 B.C. Aristarchos of Samos suggested that only the Moon revolved about the Earth. The planets, including Earth itself, he said, revolved about the Sun, thus elaborating a heliocentric system. He even had some good notions concerning the relative sizes and distances of the Moon and the Sun.

For a while, the Aristarchean view seemed to have an outside chance despite the great prestige of Aristotle. However, about 150 B.C., Hipparchos of Nicaea worked out the mathematics of the geocentric system so thoroughly that the competition ended. About 150 A.D. Claudius Ptolemy put the final touches on the geocentric theory and no one questioned that the Earth was the center of the universe for nearly 1400 years thereafter.

But had Prometheus and Vulturius been in the sky, the Greeks would have had an example of one heavenly body, anyway, that clearly did not revolve primarily about the Earth. Vulturius would have revolved about Prometheus.

Aristarchos would undoubtedly have suggested Prometheus to be another sun with a planet circling it. The argument by analogy would, it seems to me, certainly have won out. Copernicus would have been anticipated.

Furthermore, the motions of Vulturius about Prometheus would have given a clear indication of the workings of gravity. The Aristotelian notion that gravity was confined to Earth alone and that heavenly bodies were immune to it would not have stood up.

Undoubtedly Newton too would have been anticipated by some two thousand years.

What would have happened next? Would Greek genius have decayed anyway? Would the Dark Ages still have intervened? Or would the world have had a two thousand year head start in science and would we now be masters of space? Or would we possibly be the non-survivors of a nuclear war fought in Roman times?

So that's how it goes. You start off checking on colored shadows in a science-fiction story and end up wondering how different human history might have been (either for good or evil) if only the Sun had had a companion star in its lonely voyage through eternity.

PART III / THE UNIVERSE

10 *Heaven on Earth*

> *The nicest thing about writing these essays is the* constant mental exercise it gives me. Unceasingly, I must keep my eyes and ears open for anything that will spark something that will, in my opinion, be of interest to the reader.

For instance, a letter arrived today, asking about the duodecimal system, where one counts by twelves rather than by tens, and this set up a mental chain reaction that ended in astronomy and, what's more, gave me a notion which, as far as I know, is original with me. Here's how it happened.

My first thought was that, after all, the duodecimal system *is* used in odd corners. For instance, we say that 12 objects make 1 dozen and 12 dozen make 1 gross. However, as far as I know, 12 has never been used as the base for a number system, except by mathematicians in play.

A number which has, on the other hand, been used as the base for a formal positional notation is 60. The ancient Babylonians used 10 as a base just as we do, but frequently used 60 as an alternate base. In a number based on 60, what we call the units column would contain any number from 1 to 59, while what we call the tens column would be the "sixties" column, and our hundreds column (ten times ten) would be the "thirty-six hundred" column (sixty times sixty).

Thus, when we write a number, 123, what it really stands for is $(1 \times 10^2) + (2 \times 10^1) + (3 \times 10^0)$. And since 10^2 equals 100, 10^1 equals 10 and 10^0 equals 1, the total is $100 + 20 + 3$ or, as aforesaid, 123.

But if the Babylonians wrote the equivalent of 123, using 60 as the base, it would mean $(1 \times 60^2) + (2 \times 60^1) + (3 \times 60^0)$. And since 60^2 equals 3600, 60^1 equals 60 and 60^0 equals 1, this works out to $3600 + 120 + 3$, or 3723 by our decimal notation. Using a positional notation with the base 60 is a "sexagesimal notation" from the Latin word for sixtieth.

As the word "sixtieth" suggests, the sexagesimal notation can be carried into fractions too.

Our own decimal notation will allow us to use a figure such as 0.156, where what is really meant is $0 + \frac{1}{10} + \frac{5}{100} + \frac{6}{1000}$. The denominators, you see, go up the scale in multiples of 10. In the sexagesimal scale, the denominators would go up the scale in multiples of 60 and 0.156 would represent $0 + \frac{1}{60} + \frac{5}{3600} + \frac{6}{216,000}$, since 3600 equals 60×60, 216,000 equals $60 \times 60 \times 60$, and so on.

Those of you who know all about exponential notation will no doubt be smugly aware that $\frac{1}{10}$ can be written 10^{-1}, $\frac{1}{100}$ can be written 10^{-2} and so on, while $\frac{1}{60}$ can be written 60^{-1}, $\frac{1}{3600}$ can be written 60^{-2} and so on. Consequently, a full number expressed in sexagesimal notation would be something like this: $(15)(45)(2).(17)(25)(59)$, or $(15 \times 60^2) + (45 \times 60^1) + (2 \times 60^0) + (17 \times 60^{-1}) + (25 \times 60^{-2}) + (59 \times 60^{-3})$, and if you want to amuse yourself by working out the equivalent in ordinary decimal notation, please do. As for me, I'm chickening out right now.

All this would be of purely academic interest, if it weren't for the fact that we still utilize sexagesimal notation in at least two important ways, which date back to the Greeks.

The Greeks had a tendency to pick up the number 60 from the Babylonians as a base, where computations were complicated, since so many numbers go evenly into 60 that fractions are avoided as often as possible (and who wouldn't avoid fractions as often as possible?).

One theory, for instance, is that the Greeks divided the radius of a circle into 60 equal parts so that in dealing with half a radius, or a third, or a fourth, or a fifth, or a sixth, or a tenth (and so on) they could always express it as a whole number of sixtieths. Then, since in ancient days the value of π (pi) was often set equal to a rough and ready 3, and since the length of the circumference of a circle is equal to twice π times the radius, the length of that circumference is equal to 6 times the radius or to 360 sixtieths of a radius. Thus (perhaps) began the custom of dividing a circle into 360 equal parts.

Another possible reason for doing so rests with the fact that the sun completes its circuit of the stars in a little over 365 days, so that in each day it moves about $\frac{1}{365}$ of the way around the sky. Well, the ancients weren't going to quibble about a few days here and there and 360 is so much easier

Von Guericke's "Magdeburg hemispheres" experiment, 1650: a scientific-minded mayor proves that air has weight. *(Thin Air)*   THE BETTMANN ARCHIVE

The Aurora Borealis in Alaska: thin wisps of gas glow after bombardment with particles from outer space. *(Thin Air)*

THE AMERICAN MUSEUM OF NATURAL HISTORY

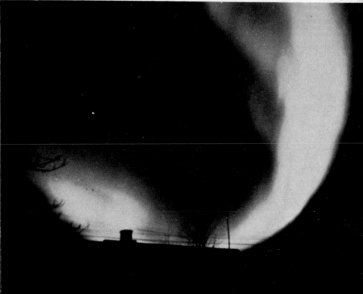

The first balloonists go up—the Montgolfier brothers making the first manned ascent, 1783. *(Thin Air)*

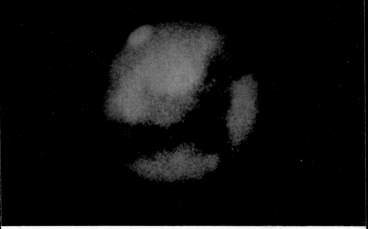

Mars: a "summer resort" of the 21st century?
(Catskills in the Sky)   YERKES OBSERVATORY

Saturn and its rings: "a uniquely beautiful sight"—
but getting a good viewing angle is a problem!
(Catskills in the Sky)   YERKES OBSERVATORY

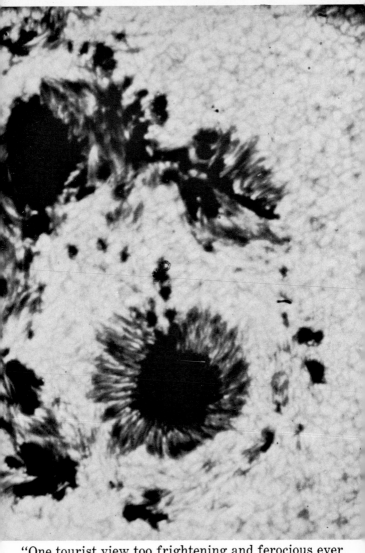

"One tourist view too frightening and ferocious ever to be really popular... a close look at the Sun." *(Catskills in the Sky)* SPECTROHELIOGRAPH OF THE SUN'S SURFACE FROM NATIONAL SCIENCE FOUNDATION

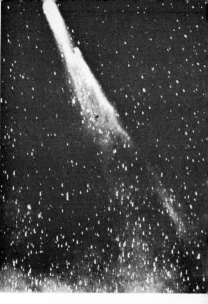

Halley's Comet: "The last halfway spectacular comet showed up in 1910, and succeeded in frightening many people into believing the end of the world had come." *(Steppingstones to the Stars)*

YERKES OBSERVATORY

The crowded center of our Galaxy. *(Our Lonely Planet)*   YERKES OBSERVATORY

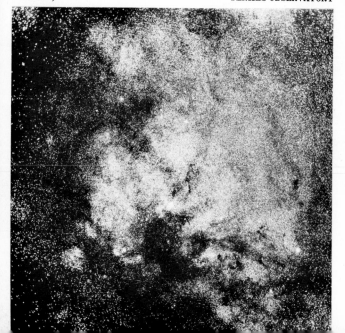

Nebula M33 in *Triangulum:* "The spiral arms of galaxies are loaded with dust out of which stars are forming or within which they are growing."
(Our Lonely Planet)

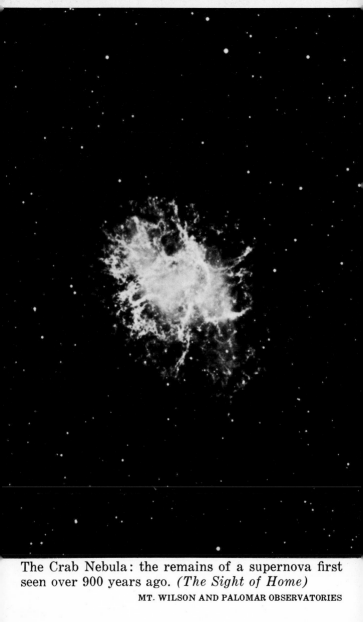

The Crab Nebula: the remains of a supernova first
seen over 900 years ago. *(The Sight of Home)*

Galileo: his doubts made him miss out—once.
(My Built-in Doubter)   THE BETTMANN ARCHIVE

to work with that they divided the circuit of the sky into that many divisions and considered the sun as traveling through one of those parts (well, just about) each day.

A 360th of a circle is called a "degree" from Latin words meaning "step down." If the sun is viewed as traveling down a long circular stairway, it takes one step down (well, just about) each day.

Each degree, if we stick with the sexagesimal system, can be divided into 60 smaller parts and each of those smaller parts into 60 still smaller parts and so on. The first division was called in Latin _pars minuta prima_ (first small part) and the second was called _pars minuta secunda_ (second small part), which have been shortened in English to minutes and seconds respectively.

We symbolize the degree by a little circle (naturally), the minute by a single stroke, and the second by a double stroke, so that when we say that the latitude of a particular spot on earth is 39° 17′ 42″, we are saying that its distance from the equator is 39 degrees plus $^{17}/_{60}$ of a degree plus $^{42}/_{3600}$ of a degree, and isn't that the sexagesimal system?

The second place where sexagesimals are still used is in measuring time (which was originally based on the movements of heavenly bodies). Thus we divide the hour into minutes and seconds and when we speak of a duration of 1 hour, 44 minutes, and 20 seconds, we are speaking of a duration of 1 hour plus $^{44}/_{60}$ of an hour plus $^{20}/_{3600}$ of an hour.

You can carry the system further than the second and, in the Middle Ages, Arabic astronomers often did. There is a record of one who divided one sexagesimal fraction into another and carried out the quotient to ten sexagesimal places, which is the equivalent of 17 decimal places.

Now let's take sexagesimal fractions for granted, and let's consider next the value of breaking up circumferences of circles into a fixed number of pieces. And, in particular, consider the circle of the ecliptic along which the sun, moon, and planets trace their path in the sky.

After all, how _does_ one go about measuring a distance along the sky? One can't very well reach up with a tape measure. Instead the system, essentially, is to draw imaginary lines from the two ends of the distance traversed along the ecliptic (or along any other circular arc, actually) to the center of the circle, where we can imagine our eye to be, and to measure the angle made by those two lines.

The value of this system is hard to explain without a diagram, but I shall try to do so, with my usual dauntless bravery (though you're welcome to draw one as I go along, just in case I turn out to be hopelessly confusing).

Suppose you have a circle with a diameter of 115 feet, and another circle drawn about the same center with a diameter of 230 feet, and still another drawn about the same center with a diameter of 345 feet. (These are "concentric circles" and would look like a target.)

The circumference of the innermost circle would be about 360 feet, that of the middle one 720 feet and that of the outermost 1080 feet.

Now mark off 1/360 of the innermost circle's circumference, a length of arc 1 foot long, and from the two ends of the arc draw lines to the center. Since 1/360 of the circumference is 1 degree, the angle formed at the center may be called 1 degree also (particularly since 360 such arcs will fill the entire circumference and 360 such central angles will consequently fill the entire space about the center).

If the 1-degree angle is now extended outwards so that the arms cut across the two outer circles, they will subtend a 2-foot arc of the middle circle and a 3-foot circle of the outer one. The arms diverge just enough to match the expanding circumference. The lengths of the arc will be different, but the fraction of the circle subtended will be the same. A 1-degree angle with vertex at the center of a circle will subtend a 1-degree arc of the circumference of any circle, regardless of its diameter, whether it is the circle bounding a proton or bounding the Universe (if we assume a Euclidean geometry, I quickly add). The same is analogously true for an angle of any size.

Suppose your eye was at the center of a circle that had two marks upon it. The two marks are separated by 1/6 the circumference of the circle, that is by 360/6 or 60 degrees of arc. If you imagine a line drawn from the two marks to your eye, the lines will form an angle of 60 degrees. If you look first at one mark, then at the other you will have to swivel your eyes through an angle of 60 degrees.

And it wouldn't matter, you see, whether the circle was a mile from your eye or a trillion miles. If the two marks were 1/6 of the circumference apart, they would be 60 degrees apart, regardless of distance. How nice to use such a measure, then, when you haven't the faintest idea of how far away the circle is.

So, since through most of man's history astronomers had

no notion of the distance of the heavenly objects in the sky, angular measure was just the thing.

And if you think it isn't, try making use of linear measure. The average person, asked to estimate the diameter of the full moon *in appearance*, almost instinctively makes use of linear measure. He is liable to reply, judiciously, "Oh, about a foot."

But as soon as he makes use of linear measure, he is setting a specific distance, whether he knows it or not. For an object a foot across to look as large as the full moon, it would have to be 36 yards away. I doubt that anyone who judges the moon to be a foot wide will also judge it to be no more than 36 yards distant.

If we stick to angular measure and say that the average width of the full moon is 31' (minutes), we are making no judgments as to distance and are safe.

But if we're going to insist on using angular measure, with which the general population is unacquainted, it becomes necessary to find some way of making it clear to everyone. The most common way of doing this, and to picture the moon's size, for instance, is to take some common circle with which we are all acquainted and calculate the distance at which it must be held to look as large as the moon.

One such circle is that of the twenty-five-cent piece. Its diameter is about 0.96 inches and we won't be far off if we consider it just an inch in diameter. If a quarter is held 9 feet from the eye, it will subtend an arc of 31 minutes. That means it will look just as large as the full moon does, and, if it is held at that distance between your eye and the full moon, it will just cover it.

Now if you've never thought of this, you will undoubtedly be surprised that a quarter at 9 feet (which you must imagine will look quite small) can overlap the full moon (which you probably think of as quite large). To which I can only say: Try the experiment!

Well, this sort of thing will do for the sun and the moon but these, after all, are, of all the heavenly bodies, the largest in appearance. In fact, they're the only ones (barring an occasional comet) that show a visible disc. All other objects are measured in fractions of a minute or even fractions of a second.

It is easy enough to continue the principle of comparison by saying that a particular planet or star has the apparent diameter of a quarter held at a distance of a mile or ten miles or a hundred miles and this is, in fact, what is generally done. But of what use is that? You can't see a quarter at all, at such distances, and you can't picture its size. You're just substituting one unvisionable measure for another.

There must be some better way of doing it.

And at this point in my thoughts, I had my original (I hope) idea.

Suppose that the earth were exactly the size it is but were a huge, hollow, smooth, transparent sphere. And suppose you were viewing the skies not from earth's surface, but with your eye precisely at earth's center. You would then see all the heavenly objects projected onto the sphere of the earth.

In effect, it would be as though you were using the entire globe of the earth as a background on which to paint a replica of the celestial sphere.

The value of this is that the terrestrial globe is the one sphere upon which we can easily picture angular measurement, since we have all learned about latitude and longitude which *are* angular measurements. On the earth's surface, 1 degree is equal to 69 miles (with minor variations, which we can ignore, because of the fact that the earth is not a perfect sphere). Consequently, 1 minute, which is equal to $\frac{1}{60}$ of a degree, is equal to 1.15 miles or to 6060 feet, and 1 second, which is equal to $\frac{1}{60}$ of a minute, is equal to 101 feet.

You see, then, that if we know the apparent angular diameter of a heavenly body, we know exactly what its diameter would be if it were drawn on the earth's surface to scale.

The moon, for instance, with an average diameter of 31 minutes by angular measure, would be drawn with a diameter of 36 miles, if painted to scale on the earth's surface. It would neatly cover all of Greater New York or the space between Boston and Worcester.

Your first impulse may be a "WHAT!" but this is not really as large as it seems. Remember, you are really viewing this scale model from the center of the earth, four thousand miles from the surface, and just ask yourself how large Greater New York would seem, seen from a distance of 4000 miles. Or look at a globe of the earth, if you have one and picture a circle with a diameter stretching from Boston to Worcester and you will see that it is small indeed compared to the whole surface of the earth, just as the moon itself is small indeed compared to the whole surface of the sky.

Actually, it would take the area of 490,000 bodies the size of the moon to fill the entire sky, and 490,000 bodies the size of our painted moon to fill all of the earth's surface.)

But at least this shows the magnifying effect of the device I am proposing, and it comes in particularly handy where bodies smaller in appearance than the sun or the moon are concerned, just at the point where the quarter-at-a-distance-of-so-many-miles notion breaks down.

For instance, in Table 1, I present the maximum angular diameters of the various planets as seen at the time of their closest approach to earth, together with their linear diameter to scale if drawn on earth's surface.

I omit Pluto because its angular diameter is not well known. However, if we assume that planet to be about the size of Mars, then at its furthest point in its orbit, it will still have an angular diameter of 0.2 seconds and can be presented as a circle 20 feet in diameter.

TABLE 1 Planets to Scale

Planet	Angular diameter (SECONDS)	Linear diameter (FEET)
Mercury	12.7	1280
Venus	64.5	6510
Mars	25.1	2540
Jupiter	50.0	5050
Saturn	20.6	2080
Uranus	4.2	425
Neptune	2.4	240

Each planet could have its satellites drawn to scale with great convenience. For instance, the four large satellites of Jupiter would be circles ranging from 110 to 185 feet in diameter, set at distances of 3 to 14 miles from Jupiter. The entire Jovian system to the orbit of its outermost satellite (Jupiter IX, a circle about 5 inches in diameter) would cover a circle about 350 miles in diameter.

The real interest in such a setup, however, would be the stars. The stars, like the planets, do not have a visible disc to the eyes. Unlike the planets, however, they do not have a visible disc even to the largest telescope. The planets (all but Pluto) can be blown up to discs even by moderate-sized telescopes; not so the stars.

By indirect methods the apparent angular diameter of some stars has been determined. For instance, the largest angular diameter of any star is probably that of Betelgeuse, which is 0.047 seconds. Even the huge 200-inch telescope cannot magnify that diameter more than a thousandfold, and under such magnification the largest star is still less than 1 minute of arc in appearance and is therefore no more of a disc to the 200-incher than Jupiter is to the unaided eye. And of course, most stars are far smaller in appearance than is giant Betelgeuse. (Even stars that are in actuality larger than Betelgeuse are so far away as to appear smaller.)

But on my earth scale, Betelgeuse with an apparent diameter of 0.047 seconds of arc would be represented by a circle about 4.7 feet in diameter. (Compare that with the 20 feet of even distantest Pluto.)

However, it's no use trying to get actual figures on angular diameters because these have been measured for very few stars. Instead, let's make the assumption that all the stars have the same intrinsic brightness the sun has. (This is not so, of course, but the sun is an average star, and so the assumption won't radically change the appearance of the universe.)

Now then, area for area, the sun (or any star) remains at constant brightness to the eye regardless of distance. If the sun were moved out to twice its present distance, its apparent brightness would decrease by four times but so would its apparent surface area. What we could see of its area would be just as bright as it ever was; there would be less of it, that's all.

The same is true the other way, too. Mercury, at its closest approach to the sun, sees a sun that is no brighter per square second than ours is, but it sees one with ten times as many square seconds as ours has, so that Mercury's sun is ten times as bright as ours.

Well, then, if all the stars were as luminous as the sun, then the apparent area would be directly proportional to the apparent brightness. We know the magnitude of the sun (−26.72) as well as the magnitudes of the various stars, and that gives us our scale of comparative brightness, from which we can work out a scale of comparative areas and, therefore, comparative diameters. Furthermore, since we know the angular measure of the sun, we can use the comparative diameters to calculate the comparative angular measures which, of course, we can convert to linear diameters (to scale) on the earth.

But never mind the details (you've probably skipped the previous paragraph already), I'll give you the results in Table 2.

(The fact that Betelgeuse has an apparent diameter of 0.047 and yet is no brighter than Altair is due to the fact that Betelgeuse, a red giant, has a lower temperature than the sun and is much dimmer per unit area in consequence. Remember that Table 2 is based on the assumption that all stars are as luminous as the sun.)

So you see what happens once we leave the Solar System. Within that system, we have bodies that must be drawn to scale in yards and miles. Outside the system, we deal with bodies which, to scale, range in mere inches.

TABLE 2   Stars to Scale

Magnitude of star	Angular diameter (SECONDS)	Linear diameter (INCHES)
−1 (e.g. Sirius)	0.014	17.0
0 (e.g. Rigel)	0.0086	10.5
1 (e.g. Altair)	0.0055	6.7
2 (e.g. Polaris)	0.0035	4.25
3	0.0022	2.67
4	0.0014	1.70
5	0.00086	1.05
6	0.00055	0.67

If you imagine such small patches of earth's surface, as seen from the earth's center, I think you will get a new vision as to how small the stars are in appearance and why telescopes cannot make visible discs of them.

The total number of stars visible to the naked eye is about 6000, of which two thirds are dim stars of 5th and 6th magnitude. We might then picture the earth as spotted with 6000 stars, most of them being about an inch in diameter. There would be only a very occasional larger one; only 20, all told, that would be as much as 6 inches in diameter.

The average distance between two stars on earth's surface would be 180 miles. There would be one or, at most, two stars in New York State, and one hundred stars, more or less, within the territory of the United States (including Alaska).

The sky, you see, is quite uncrowded, regardless of its appearance.

Of course, these are only the visible stars. Through a telescope, myriads of stars too faint to be seen by the naked eye can be made out and the 200-inch telescope can photograph stars as dim as the 22nd magnitude.

A star of magnitude 22, drawn on the earth to scale, would be a mere 0.0004 inches in diameter, or about the size of a bacterium. (Seeing a shining bacterium on earth's surface from a vantage point at earth's center, 4000 miles down, is a dramatic indication of the power of the modern telescope.)

The number of individual stars visible down to this magnitude would be roughly two billion. (There are, of course, at least a hundred billion stars in our Galaxy, but almost all of them are located in the Galactic nucleus which is completely hidden from our sight by dust clouds. The two billion we do see are just the scattering in our own neighborhood of the spiral arms.)

Drawn to scale on the earth, this means that among the 6000 circles we have already drawn (mostly an inch in diameter) we must place a powdering of two billion more dots, a small proportion of which are still large enough to see, but most of which are microscopic in size.

The average distance between stars even after this mighty powdering would still be, on the earth-scale, 1700 feet.

This answers a question I, for one, have asked myself in the past. Once a person looks at a photograph showing the myriad stars visible to a large telescope, he can't help wondering how it is possible to see beyond all that talcum powder and observe the outer galaxies.

Well, you see, despite the vast numbers of stars, the clear space between them is still comparatively huge. In fact, it has been estimated that all the starlight that reaches us is equivalent to the light of 1100 stars of magnitude 1. This means that if all the stars that can be seen were massed together, they would fill a circle (on earth-scale) that would be 18.5 feet in diameter.

From this we can conclude that all the stars combined do not cover up as much of the sky as the planet Pluto. As a matter of fact, the moon, all by itself, obscures nearly 300 times as much of the sky as do all other nighttime heavenly bodies, planets, satellites, planetoids, stars, put together.

There would be no trouble whatever in viewing the spaces outside our Galaxy if it weren't for the dust clouds. Those are really the only obstacle, and they can't be removed even if we set up a telescope in space.

What a pity the universe couldn't really be projected on earth's surface temporarily—just long enough to send out the Walrus's seven maids with seven mops with strict orders to give the universe a thorough dusting.

How happy astronomers would then be!

11 Our Lonely Planet

One of the questions that is asked innumerable times these daring days (I have even asked it myself) is: "If there is life elsewhere in the Universe, why hasn't *it* reached *us?*"

Since modern views of the Universe would have it that solar systems are the rule rather than the very rare exception we thought they were twenty years ago, there should be millions, perhaps billions, of planets with physical and chemical characteristics reasonably close to those of Earth in our own Galaxy alone. Since modern views on biochemistry would seem to make the origin of life the inevitable consequence of Earthlike physics and chemistry rather than a rare and miraculous occurrence, there should also be millions, perhaps billions, of independent life-systems in our Galaxy alone.

Since most other planets are probably as old as our own, there has been ample time for evolution elsewhere as well as here. Suppose that one system of planetary life out of a thousand develops organisms with sufficient intelligence to understand and control the forces of Nature. Then there are still thousands, perhaps millions, of intelligent life-forms in our own Galaxy alone.

Now, to repeat: "If there is life elsewhere in the Universe, why hasn't it reached *us?*"

Well, through a circuitous line of reasoning, I think I have a possible answer that sounds well to me. The line of reasoning starts in the Bible.

In Genesis 15:5, it is related that the Lord encouraged the patriarch Abram, who feared that, since he was childless, earlier promises that he would be made "a great nation" would not, after all, be kept. The verse reads: "And he [the Lord] brought him forth abroad and said, Look now toward heaven, and tell the stars, if thou be able to number them: and he said unto him, So shall thy seed be."

This is an example of the typical way in which the ancients expressed large numbers: "as the stars in the heaven" or "as the sand grains of the shore" or "as the water drops of the ocean."

Now there *are* many drops of water in the ocean and many grains of sand in the sea shore. As far as ancient man was concerned, such numbers *were* infinite. Except for occasional geniuses like Archimedes, no man before modern times could even consider that numbers might exist great enough to express sand grains and water drops. (It was 1300 A.D. before the word "million" was invented. Until then, the largest number word was "myriad," which was Greek for 10,000. Even Archimedes, in calculating the number of poppy seeds in the entire Universe as he knew it, used expressions meaning "myriads of myriads of myriads. . . .")

But what about the number of stars in the sky? Are they as innumerable as the sand grains and water drops?

To be sure, if the Lord had chosen to reveal to Abram *all* the stars in the Universe in one flash of miraculous sight, Abram would have seen a minimum of 10,000,000,000,000,-000,000,000 (ten billion trillion) stars, and that would certainly have been innumerable to him.

However, no Biblical commentator I've ever heard of has suggested that this was what happened. The suggestion of vast numbers in Genesis 15:5 is always taken as applying to only those stars actually visible in the sky to the naked eye.

Even with that limitation, most people, I'm sure, would consider the metaphor quite an effective one and not in the least ridiculous.

I can see why they should think that, too. I myself am a city boy and I've hardly ever really seen the stars. The buildings block them out; the street lights dim them out; smoke and dirt blot them out. And only once did I get a real chance to find out what I was missing.

I spent the night at a friend's country house in New Hampshire, you see, and when night came, I couldn't sleep. It got dark, it seems. Battling my primitive fears of a darkness I had never really experienced, I thought I would step out into the open and prove to myself that there was nothing to be afraid of. It was a warm summer night, so I just walked outside in my pajamas and slippers.

There was no Moon, no clouds, no artificial light for miles around. So for the first time in my life, *I saw the stars!*—millions of them, billions of them, trillions of them.

It was a wonderful sight. I stayed out, staring, for a long time, and to the day I die I shall always remember that once I saw the stars.

But the question is, how many stars did I really see?

The faintest star that can be seen with the naked eye, under the best conditions, is of magnitude 6.5 and the number of stars that exist in the entire circuit of the skies that bright or brighter is just about 6000.

That's all. That's the hard fact of it. Six thousand.

And since at any moment only half the sky is above the horizon, the number of stars theoretically visible at any one moment is 3000. But then the atmosphere absorbs some of the light that passes through it. Even the purest, clearest atmosphere absorbs 30 per cent of the starlight that strikes it. As you turn your gaze toward the horizon, you are looking through a much greater thickness of air than you do when you stare up at zenith. The result is that the faintest stars which can be just made out near the zenith are lost to vision if they are located lower in the sky.

Actually, then, the total number of stars I could possibly have seen outside my friend's summer house (even counting those obscured by trees and unevenness of the horizon) was 2500.

The stars in the sky innumerable? Hah. Even the Babylonian shepherds could have counted to 2500, I'm sure.

One dramatic way of pointing out the difference between the facts as they are and the facts as we think they are is to pose the following riddle: If, at any particular time, the Moon were removed from the sky, how many stars (visible ones, of course) would it have been covering?

If one thinks of the size of the Moon and the thickness with which stars are strewn across the vault of the night sky and then follows intuition, the answer given might be five or seven or ten or fifty.

Anyway, quite a few. What's *your* guess?

But let's not go by intuition. The circuit of the skies is measured by degrees—360 degrees for the full circumference. The area of the total sky (or of any sphere, for that matter) works out to about 41,200 square degrees. Since there are

6000 visible stars all told, we can say that there is one star for every 6.9 square degrees of sky.

But the apparent diameter of the Moon's sphere is (on the average) 0.52 degrees. Its area is therefore 0.21 square degrees and the odds are thus about 33 to 1 that the removal of the Moon would reveal *not one star behind it*.

This stars-in-the-sky situation changes at once if we view the skies from the Moon, or from a space station, or from any point outside a planetary atmosphere. Science-fiction writers usually talk about the "familiar constellations" seen from other worlds in our Solar System. Yet the notion is almost certainly wrong.

The reasoning behind the "familiar constellations" business is that any change of position within the Solar System involves so small a movement in comparison with the distances of the stars that there would be no noticeable alteration in their relative positions.

And that's right, as far as it goes.

However, remember the 30 per cent of starlight that is absorbed by our atmosphere. On the Moon, to use that as an example, no starlight is absorbed and every individual star seems 1¾ times as bright as it does to us on Earth. Another way of putting it is that every star is 0.4 lower (i.e. brighter) in magnitude on the Moon than on the Earth.

This is a noticeable increase in brightness but not an overwhelming one. The eye would grow accustomed to it quickly and, if that were all, the Moon's starry sky would seem gaudy (with its brighter and non-twinkling stars) but not strange.

But it's not all. Allow for this uniform increase of 0.4 magnitude and the limit of naked-eye visibility stretches down to stars of magnitude 6.9. That is, a star which is of magnitude 6.9 on Earth (and therefore invisible to the naked eye) is of magnitude 6.5 as seen from the Moon and just visible.

So what?

So this: The number of stars increases very rapidly with magnitude. Any glance at the sky will convince you that there are many more dim stars than bright stars. To be bright, a star has to be big or close. Well, there are more small stars than big ones and since volume increases as the cube of the radius, there is more room far away than close by. In general, the number of stars at each level of magnitude is three times the number at the previous level. Thus there are about 350 stars between magnitudes 3 and 4, about 1100 between magnitudes 4 and 5, and about 3200 between magnitudes 5 and 6.

In the interval between 6.5 and 6.9 there are 6000 stars. All of these are not visible on Earth and are visible on the Moon, just because the Moon lacks an atmosphere. The night sky as seen from the Moon, therefore contains 12,000 stars, *twice the number that can be seen from Earth.* Furthermore the number that can be seen above the horizon at any one time is not lessened by the effect of additional atmospheric absorption. The actual number seen at any one time from level ground on the Moon is therefore 2½ times the number seen under similar conditions on Earth.

From the Moon (or in space, generally) you could still make out patterns of bright stars such as those of the Big Dipper or of Orion, but the finer details would all be drowned out in thousands of additional stars, and the over-all effect would be that of a completely strange sky.

In other words, when we leave Earth we say farewell to our dear "familiar constellations."

This raises another point. Are there places in the Universe where the starry sky is even more impressive than it appears from the Moon?

Obviously, it would be more impressive to inhabitants who lived on a planet revolving about a Sun that was part of the densely populated central nucleus of a Galaxy or within a globular cluster. Our own Sun, after all, is way out in the sparsely-populated spiral arm of our Galaxy.

In our home neighborhood there are 188 stars or star systems (that is, binaries or multiple stars) known to be located within 10 parsecs of Earth. (A parsec is equal to 3.26 light-years). This means that, on the average, there are 4½ stars (or star systems) per 100 cubic parsecs of space and that the average distance between stars (or star systems) in our neck of the woods is about 2.8 parsecs, which is equivalent to 9.2 light-years.

At the Galactic center or in a globular cluster (a positive photograph of which, under high magnification, looks for all the world like a heap of talcum powder) the average distance between stars is one light-year. A hundred cubic parsec volumes in which stars were this closely packed would contain not 4½ stars but 3500 stars.

In other words, all things being otherwise equal, the number of stars visible in the skies near the Galactic center would be 780 times as many as those visible out here. Even allowing for horizon effects, th number of stars visible above the horizon would be abo t 2,000,000.

There would be, on the averag , 100 visible stars per square degree of the sky and a globe the area of the Moon would be covering 20 stars, on the average.

There would naturally be more stars at every level of brightness. The skies in the Galactic center would contain more first magnitude stars (about 7500) than our heavens contain of stars of any description.

Furthermore, the chances are very much in favor of there being a number of stars brighter than any of those in our own skies. We can duplicate Galactic center conditions by imagining all the visible stars we see pulled in to 1/9.2 of their actual distance. Any star whose nearness is increased 9.2 times has its brightness increased 9.2 × 9.2 or 85 times. A brightness increase of 85 is equivalent to a decrease in magnitude of 4.8.

Sirius, in other words, instead of the magnitude of −1.6 which it now has, would burn with a brightness equal to a magnitude of −6.4. It would be eight times as bright as Venus at its brightest. Ten other stars in our sky would become brighter than Venus under these conditions and about 250 stars altogether would be brighter than Sirius (our brightest star) appears to us now.

The starlight in such a sky would by no means be negligible. It would be roughly equal to the light of the full Moon as seen from Earth, so that a cloudless night, under such conditions, would never be dark.

Despite all this gorgeous display, the stars would still all look like stars. The chances of having any stars close enough to look like tiny suns with visible globes is just about nil.

Assuming our Sun to be an average star and placing it at a light-year's distance (the average interstellar distance at the Galactic center), its apparent diameter would be about 0.03 seconds of arc. (There are 60 seconds to 1 minute and 60 minutes to 1 degree.) In order for a heavenly body to be seen as a globe it must have an apparent diameter of at least 3 minutes. Even the 200-inch Palomar telescope would not show the Sun to be a tiny globe at a distance of a light-year.

Of course, one light-year is only the average distance between stars. Some stars would be closer to one another. Well, in order for a star the size of the Sun to be seen as a globe it would have to be distant not more than a billion miles, which is less than the distance between us and the planet

Uranus. It is quite impossible for a star to be that close to another unless it is part of a binary, and I'm not considering that situation here.

But then again, suppose the star to be larger than the Sun. All right. In order for a star to be seen as a globe at a distance of a light-year, it would have to have a diameter of about 8000 times that of the Sun. If such a star were in the position of our Sun, it would fill up the Solar System beyond the orbit of Neptune. Stars that size are freaks indeed and the chances of finding one within a light-year of your planet are virtually nil.

Now what does all this have to do with the neglect of our own world by the possible intelligences elsewhere in our Galaxy? Consider several points:

1. About 90 per cent of the stars and hence, assuming random distribution, about 90 per cent of the intelligences which have evolved, exist in the crowded Galactic center.

2. A closer spacing of stars makes interstellar travel somewhat less of a problem while the concomitantly greater "starriness" of the sky is liable to make interstellar travel more of a popular goal and dream.

3. The intermingling of cultures is a catalyst for advancement.

Now then, if all intelligences have an equal chance of being the first to attain interstellar travel, then on the basis of point 1, it is nine times as likely that the victory be attained first somewhere in the Galactic center.

If the chances are not equal but are inversely proportional to the average distance separating the stars, then, combining points 1 and 2, it is eighty times as likely that the victory be attained first in the Galactic center.

Once one group achieves interstellar travel, other intelligences that are reached by it will either be wiped out or colonialized, or will also learn the trick and proceed to spread it to those intelligences *they* can reach. Therefore what I mean by point 3 is that although it might take six billion years for one world to develop a life-form that can, in turn, develop interstellar travel; it will then perhaps take as little as a thousand years for all the intelligences within reach to develop it also.

In short, then, if even one group of intelligences has worked up a practical form of interstellar travel during the last few millennia, I find no trouble in imagining that trade has already developed on a Galactic scale or that even some kind of Galactic federation already exists. (Perhaps there are some small independent federations, each not knowing of the other's existence, among the various globular clusters.)

But then why hasn't the federation contacted us?

Easy. I used the phrase "intelligences within reach" a few lines back and that's the key point.

Consider the economics of the thing. With 90 per cent of the worlds, of the resources, of the intelligences in the Galactic center, why bother venturing out into the spiral arms, where distances that must be covered between stars are nine times as great and the pickings in terms of worlds, of resources, of intelligences is only one tenth as great?

When a sample of iron ore is too low in iron, it becomes unprofitable to work it. And when a sample of space becomes too thin in worlds, is it too unprofitable to enter it?

If so, here we are on our lonely planet, a bunch of hicks way out in the sticks, lost in the backwoods where no reasonable beings would waste the energy to go. And if so, that's the way we're likely to stay, too, unless we find methods of spanning interstellar distances ourselves, go down to the Big Town we call the Galactic center and force ourselves on the city slickers.

Maybe we'll do just that someday—if all this is so.

But is all this so? More particularly, is it true that mere distance need be such a barrier. It's a natural tendency to consider the light-speed limit absolute and to think of interstellar travel as involving years, centuries, millennia, and not to think of intergalactic travel at all. Yet must we think in this manner?

Down to 1800 we knew of no way in which a man could move more quickly than a horse's straining muscles could carry him, or than a gale could force a ship through water. It didn't prevent the imaginative fiction writers of those days from thinking up devices such as flying horses, flying carpets, seven-league boots, and obliging demons. None of that ever came to pass; it was all gibberish. But locomotives, autos, planes, and jets came to pass and, really, those did the trick more efficiently, more reliably, or both.

The imaginative fiction writers of these days try to get around the light-speed limit by thinking up devices such as hyperspace, inertia-less drive and so on. That is just gibberish, too, and perhaps as unlikely as flying carpets. Nevertheless, the real equivalent may someday exist and distance may become quite unimportant as a barrier. (On the surface of

Earth, today, show people commute between the East and West Coasts, and it matters little to us that a prime source of uranium is in the Congo. Think how either situation would have struck us as little as fifty years ago.)

Now then, suppose distance *doesn't* matter in a Galactic Empire, and that our little speck of life way out here is just about as accessible as any world in the center. Is there any remaining reason why alien life hasn't reached us?

Yes—chance!

Just by the breaks of the game, they haven't happened to reach us yet.

Let's look into that, now. A question of age arises. Since man has begun wondering about fossils and the length of time it took to put down a particular thickness of sedimentary rock, he has been lengthening the age of the Universe constantly.

As late as ten years ago, the accepted age of the Universe was a mere two billion years. Then it turned out that some calculations on Cepheid variables were wrong; the Galaxies were twice as far apart as had been thought and the Universe was four billion years old. The latest figures I have heard is that the Sun is five billion years old and the Universe at least twenty-four billion years old.

The end is not yet. We still don't know enough.

But some things we do know. For instance, some stars are considerably younger than others; stars were *not* all formed at the same time.

For one thing, the giant hot stars (spectral classes O and B) are expending energy at so fearsome a rate that they can't possibly have a lifetime of more than some millions of years—a lifetime, that is, of active starhood prior to the white-dwarf stage.

There are also places where cosmic dust spreads thickly, within which it is thought stars are actually forming now.

In fact, the cosmic dust holds the key. The spiral arms of Galaxies (most definitely including our own) are loaded with dust out of which stars are forming or within which they are growing. The most dust, the more instability. Our own immediate neighborhood is relatively dust-free and the Sun is a sober, respectable star of settled habits and considerable age.

The stars in the spiral arms of Galaxies are included in "Population I."

In the Galactic center (of this and other Galaxies) and in globular clusters, space is clear, however. There is no dust to speak of. The result is that the stars located there are "Population II" stars. They are quiet stars, of approximately equal age and condition, not growing or undergoing spectacular change. In general, Population II stars seem to be older than Population I stars—how much older has not gone beyond the realm of speculation yet, but a seniority of several billion years (on the average) for Population II over Population I is a possibility.

One could imagine that in the initial ages of Galaxy formation, stars formed rapidly and by the tens of billions in the various Galactic centers (around a main nucleus that formed the Center proper, and around subsidiary small nuclei that formed a halo of globular clusters). The thick swarming of stars quickly consumed the raw material (that is, the Galactic dust) out of which they were formed. Eventually, no more dust; only stars. Furthermore, if the distribution of dust were reasonably uniform, you would expect a reasonably uniform distribution of stars of reasonably uniform stellar characteristics.

But during the process of formation, the Galaxies were rotating. A certain amount of dust was thrown out by centrifugal forces and formed the spiral arms. The dust in the arms was, on the whole, less dense than in the Center, so that stars formed more slowly and, in fact, star formation is still proceeding. Furthermore, the dust distribution would be less uniform so that some stars would have more than an average share of dust and some less. In general, then, the stars in the arms would be of widely distributed age and stellar characteristics, but all would be, to a varying degree, younger than the stars of the center.

So now there is the possibility of supposing that the Galactic Empire had been formed billions of years ago in the old, old, old stellar systems of the center. If so, it is reasonable to assume that such an old established Empire has had the curiosity, ability, and (most important) the *time* to become aware of every life-bearing planet in the Galaxy, including ours. (Just as our own more advanced nations are certainly aware of every life-bearing speck in the South Pacific whether they are particularly interested in it or not.)

Such a consideration would remove the chance that we have not been stumbled upon by the breaks of the game, by the workings of pure chance. The alien life-forms *have* stumbled upon us.

So we reach the final question. If they *have* stumbled on us, why don't we know about it?

Maybe we can find an analogy on Earth that will help give us an answer to that.

When primitive man faced the animal kingdom, he killed every beast and bird he could; either for food or in self-defense. When he grew in sophistication and in control of his environment, he tamed certain animals, but used them for work or as a more dependable food supply. Another increase in sophistication and he kept some as pets or companions, but only the small and pleasant ones.

Nowadays in our pride of mastery and in our complete self-confidence as lords of the planet, we can afford to be completely generous. We establish zoos for even the dangerous animals and treat them kindly and well. We set up hunting seasons, protect the young, refuse to allow unlimited killing. When we find that some animal is in danger of extinction, we get all upset and do our level best to save them. (Of course, our hand is still against organisms that endanger us. I don't suppose anyone would raise a finger to prevent the extinction of the tubercle bacillus.)

This is also true on the human level. When Europeans first entered the North American continent, they wiped out such Indians as they could reach. Now the descendants of the Europeans keep the descendants of such Indian tribes as survived on reservations and feel a paternalistic responsibility for them. And by and large, it has gone out of fashion to try to wipe out "primitive" people.

So, by analogy, one can envision the youth of the Galactic Empire, in which competing intelligences might have battled viciously and ruthlessly until they learned to co-operate and tolerate, or until one particular intelligence gained undisputed sovereignty. During that period, too, any planet containing a subintelligent life-form would, if useful to one of the intelligent life-forms, be taken over with as little scruple as the Europeans displayed toward the duckbill platypuses (or even toward the aborigines) when they took over Australia.

But then once the Galactic Empire had been established and had existed for a few billion years, a change of view, born of security and maturity, might have taken place, analogous to the change of view among mankind which I've just described.

They might feel a general humanitarian protectiveness concerning the baby intelligences cropping up in the spiral arms. They might even feel curious about them. After all the stars in the arms exist through a much wider range of variation in characteristics than do the stars in the center. Perhaps as a result, the life-forms developing on their planets also exhibit a wider range of variations. The baby intelligences might show spectacular differences that would arouse intellectual interest on the part of the xenobiologists of the center.

In either case—set up the "No Trespassing" signs; "No Hunting Under Any Conditions"; run the barbed wire around the preserves; set up guards to shoot down poachers on sight; observe and make notes from afar, but don't on any account let the timid little things see you and be disturbed; and my, my, look at them explode atomic bombs just as cute and cunning as they can be—you would almost swear they were people.

So maybe we're not hicks; maybe we're protected specimens and don't know it. In which case, it's still up to us to grow up and show the big boys we're something—or at least, that we're going to be something someday.

And maybe we can do that too—some day.

12 *The Flickering Yardstick*

Every once in a while, astronomical opinion concerning the size of the Universe changes suddenly—invariably for the larger. The last time this happened, the responsibility could be placed directly at the door of a wartime blackout.

As late as the turn of the century, astronomers had only the foggiest notion of the size of the Universe, as a matter of fact. The best estimate of the time was made by a Dutch astronomer named Jacobus Cornelis Kapteyn. Beginning in 1906, he supervised a survey of the Milky Way. He would photograph small sections of it and count the stars of the various magnitudes. Assuming them to be average-sized stars, he calculated the distance they would have to be in order to show up as dimly as they did.

He ended up with the concept of the Galaxy as a lens-shaped object (something which had been more or less generally accepted since the days of William Herschel, a century earlier). The Milky Way is simply the cloudy haze formed by the millions of distant stars we see when we look through the Galactic lens the long way. Kapteyn estimated that the Galaxy was 23,000 light-years in extreme diameter and 6000 light-years thick. And as far as he, or anyone, could tell at the time, nothing existed outside the Galaxy.

He also decided that the Solar System was located quite near the center of the Galaxy, by the following line of reasoning. First, the Milky Way cut the heavens into approximately equal halves, so we must be on the median plane of the lens. If we were much above or below the median plane, the Milky Way would be crowded into one half of the sky.

Secondly, the Milky Way was about equally bright all the way around. If we were toward one end or the other of the lens, the Milky Way in the direction of the farther end would be thicker and brighter than the section in the direction of the nearer end.

In short, the Sun is in the center of the Galaxy, more or less, because the heavens are symmetrical, and there you are.

But there was one characteristic of the heavens which showed a disturbing asymmetry. Present in the sky are a number of "globular clusters." These are collections of stars packed rather tightly into a more or less spherical shape. Each globular cluster contains anywhere from a hundred thousand stars to a few million and about two hundred of them exist in our Galaxy.

Well, there's no reason why such clusters shouldn't be evenly distributed in the Galaxy, and if we were at the center, they should be spread evenly all over the sky—but they're not. A large part of them seem to be crowded together in one small part of the sky, that part covered by the constellations of Sagittarius and Scorpio.

It's the sort of odd fact that bothers astronomers and often proves the gateway to important new views of the Universe.

The way to a solution of the problem, and to the new view of the Universe, lay through a consideration of a certain kind of variable star; a star that is which is constantly varying in brightness; a star which flickers, if you like.

There are a number of different kinds of variable stars, differentiated from one another by the exact pattern of light variation. Some stars flicker for outside reasons; usually because they are eclipsed by a dim companion which gets in the way of our line of sight. The star Algol in the constellation Perseus has a dim companion which gets in our way every 69 hours. During that time of eclipse, Algol loses two thirds of its light (it is not a total eclipse) in a matter of a couple of hours and regains it as quickly.

More interesting are stars which *really* vary in brightness because of changes in their internal constitution. Some explode with greater or lesser force; some vary all over the lot in irregular fashion for mysterious reasons; and some vary in very regular fashion for equally mysterious reasons.

One of the brighter and more noticeable examples of the last group is a star called Delta Cephei, in the constellation Cepheus. It brightens, dims, brightens, dims with a period of 5.37 days. From its dimmest point it brightens steadily for about two days, reaching a peak brightness that is just double its brightness at dim point. It then spends about three and a third days fading off to its dim point again. The brightening is distinctly more rapid than the dimming.

From its spectrum, it would seem that Delta Cephei is a pulsating star. That is, it expands and contracts. If it remained the same temperature during this pulsation, it would be easy to understand that it was brightest at peak size and dimmest at least size. However, it also changes temperature and is hottest at peak brightness and coolest at the dim point. The trouble is that the peak temperature and peak brightness come, not at maximum size, but when it is expanding and halfway toward peak size. The lowest temperature and dimmest point comes when it is contracting and halfway toward minimum size. This means that Delta Cephei ends up being about the same size at the peak of brightness and in the trough of dimness. In the first case, though, it is in the process of expansion; in the latter, in the process of contraction.

Why the regular but non-synchronous pulsation in size and temperature? That part is still a mystery.

There is enough that is characteristic of Delta Cephei in all this to make astronomers realize when they found other stars behaving in like fashion, that all must belong to a group of structurally similar stars, which they called "Cepheid variables" in honor of the first of the group.

Cepheids vary among themselves in the length of their period. Some periods are as short as one day, some as long as 45 days, with examples all the way through the range. The Cepheids closest to us have periods of about a week.

The brightest and closest Cepheid is none other than the North Star. It has a period of 4 days, but during that time its flicker causes it to vary in brightness by only about 10 per cent, so it's not surprising no non-astronomer notices it, and that astronomers themselves paid more attention to the somewhat dimmer but more drastically changeable Delta Cephei.

There are a number of stars with Cepheid-like variation curves that are to be found in the globular clusters. Their main distinction from the ordinary Cepheids nearer us, is that they are extremely short-period. The longest period among these is about a day, and periods as short as an hour and a half are known. These were first called cluster Cepheids while ordinary Cepheids were called classical Cepheids. However, cluster Cepheids turned out to be a misnomer, because such stars were found with increasing frequency outside clusters, also.

The cluster Cepheids are now usually called by the name of the best-studied example (just as the Cepheids themselves are). The best-studied example is a star called RR Lyrae, so the cluster Cepheids are called RR Lyrae variables.

Now none of this seemed to have any connection with the size of the Universe until 1912 when Miss Henrietta Leavitt, studying the Small Magellanic Cloud, came across a couple of dozen Cepheids in them.

(The Large and the Small Magellanic Clouds are two foggy patches that look like detached remnants of the Milky Way. They are visible in the Southern Hemisphere and were first sighted by Europeans during the round-the-world voyage of Ferdinand Magellan and his crew back in 1520—whence the name.)

The Magellanic Clouds can be broken up into stars by a good telescope and it is only because they are a long way distant from us that these stars fade into an undifferentiated foggy patch. Because the clouds are so far distant from us, all the stars in one cloud or the other may be considered about the same distance from us. Whether a particular star is at the near edge of the cloud or the far edge makes little difference. (This is like making the equally true statement that every man in the state of Washington is roughly the same distance, i.e. 3000 miles, from an individual in Boston.)

This also means that when one star in the Small Magellanic Cloud appears twice as bright as another star, it *is* twice as bright. There is no distance difference to confuse the issue.

Well, when Miss Leavitt recorded the brightness and the period of variation of the Cepheids in the Small Magellanic Cloud, she found a smooth relationship. The brighter the Cepheid, the longer the period. She prepared a graph correlating the two and this is called the "period-luminosity curve."

Such a curve could not have been discovered from the Cepheids near us, just because of the confusing distance difference. For instance, Delta Cephei is more luminous than the North Star and therefore has a longer period. But the North Star is considerably closer to us than is Delta Cephei, so that the North Star *seems* brighter to us. For that reason, the longer period *seems* to go with the dimmer star. Of course, if we knew the actual distances of the North Star and of Delta Cephei, we could straighten the matter out, but at the time the distances were not known.

Once the period-luminosity curve was established, astronomers promptly made the assumption that it held for all

Cepheids, and were then able to make a scale model of the Universe. That is, if astronomers spotted two Cepheids with equal periods, they could assume they were also equal in actual luminosity. If Cepheid A seemed only one fourth as bright as Cepheid B, it could only be because Cepheid A was twice as distant from us as was Cepheid B. (Brightness varies inversely as the square of the distance.) Cepheids of different period could be placed, relatively to us, with only slightly more trouble.

With all the Cepheids in relative place, astronomers would need to know the actual distance in light-years to any one of them, in order to know the actual distance to all the rest.

There was only one trouble here. The sure method of determining the distance of a star was to measure its parallax. At distances of more than 100 light-years, however, the parallax becomes too small to measure. And unfortunately even the nearest Cepheid, the North Star, is several times that distance from us.

Astronomers were forced into long-drawn-out, complex, statistical analyses of medium-distant (not globular) star clusters. In this way they determined the actual distance of some of those clusters, including the Cepheids they contained.

The scale model of the Universe thus became a real map. The Cepheid variables had become flickering yardsticks in the hands of the astronomers.

In 1918 Harlow Shapley started calculating the distance of the various globular clusters from the RR Lyrae variables they contained, using Miss Leavitt's period-luminosity curve. His figures turned out to be a little too large and were corrected downward during the next decade, but the new picture of the Galaxy which grew out of his measurements has survived.

The globular clusters are distributed spherically above and below the median plane of the Galaxy. The center of this sphere of globular clusters is in the plane of the Galactic lens, but at a spot some tens of thousands of light-years from us in the direction of the constellation Sagittarius.

That explained why most of the globular clusters were to be found in that direction.

It seemed to Shapley a natural assumption that the globular clusters centered about the center of the Galaxy and later evidence from other directions bore him out. So here we are, not in the center of the Galaxy at all, but well out to one side.

We are still in the median plane of the Galaxy, for the Milky Way does split the heavens in half. But how account for the fact that the Milky Way is equally bright throughout if we are not, in fact, centered? The answer is that the median plane on the outskirts of the Galaxy (where we happen to be) is loaded with dust clouds. These happen to lie between ourselves and the Galactic center, obscuring it completely.

The result is that, with or without optical telescopes, we can only see our portion of the Galactic outskirts. We *are* in the center of that part of the Galaxy which we can see optically, and that part is not too far off in size from Kapteyn's estimate. Kapteyn's error (which was at the time excusable) was in assuming that what we could see was all the Galaxy there was. It wasn't.

The final model of the Galaxy, now thought to be correct, is that of a lens-shape that is 100,000 light-years across and 20,000 light-years thick at the center. The thickness falls off as the edge is approached and in the position of the Sun (30,000 light-years from the center and two thirds of the way toward the extreme edge of the Galaxy) is only 3000 light-years thick.

Even before the Galactic measurements had been finally determined, the Cepheids in the Magellanic Clouds had been used to determine their distance. Those turned out to be rather more than 100,000 light-years distant. (Our best modern figures are 150,000 light-years for the Large and 170,000 light-years for the Small Magellanic Cloud.) They are close enough to the body of our Galaxy and small enough in comparison, to be fairly considered "satellite Galaxies" of ours.

From the rate at which our Sun and neighboring stars are traveling in their 200,000,000-year circuit of the Galactic center, it is possible to calculate the mass of the Galactic center (which contains most of the stars in the Galaxy) and it turns out to be something like 90,000,000,000 times that of our Sun. If we consider our Sun to be an average star in mass, then we can fairly estimate the Galaxy as a whole to contain 100,000,000,000 stars. In comparison, the two Magellanic Clouds together contain a total of about 6,000,000,000.

The question in the 1920s was whether anything existed in the Universe outside our Galaxy and its satellites. Suspicion rested on certain dim, foggy structures, of which a cloudy patch in the constellation Andromeda was the most spectacular. (It was about half the size of the full moon to the naked eye and was called the Andromeda nebula—"neb-ula" coming from a Greek word for cloud.)

There were some nebulae which were known to be parts of our Galaxy because they contained hot (and not too distant) stars that were the cause of their luminosity. The Orion nebula is an example. The Andromeda nebula, however, contained no such stars that anyone could see and seemed to be self-luminous. Could it be, then, a patch of haze that could be broken up into many far, far distant stars (with the proper magnification), as could the Milky Way and the Magellanic Clouds? Since the same telescopes that resolved the Milky Way and the Magellanic Clouds did not manage to do the same for the Andromeda nebula, could it be that the Andromeda nebula was far more distant?

The answer came in 1924, when Edwin Powell Hubble turned the new 100-inch telescope at Mount Wilson on the Andromeda nebula and took photographs that showed the outskirts of the nebula resolved into stars. Furthermore, he found Cepheids among the newly revealed stars and used them to determine the distance. The Andromeda nebula turned out to be 750,000 light-years distant and this is the value found in all the astronomy books published over the next thirty years.

Allowing for the distance, the Andromeda nebula was obviously an object of galactic size, so it is now called the Andromeda galaxy. Hubble established the fact that many other nebulae of the Andromeda type were galaxies, even more distant than the Andromeda galaxy (which is a near neighbor of ours, in fact). The size of the Universe sprang instantly from a diameter in the hundreds of thousands of light-years to one in the hundreds of millions.

However, there were a few disturbing facts which lingered. For one thing, the other galaxies all seemed to be considerably smaller than our own. Why should our own Galaxy be the one outsize member of a large group?

For another, the Andromeda galaxy had a halo of globular clusters just as our own Galaxy did. Their clusters, however, were considerably smaller and dimmer than ours. Why?

Thirdly, allowing for the distance of the galaxies and the speed at which the Universe was known to be expanding it could only have been two billion years ago that all the galaxies were squashed together at some central starting point. The trouble with that was that the geologists swore up and down that the earth itself was considerably older than two billion years. How could the earth be older than the Universe?

The beginning of an answer came in 1942, when Walter Baade took another look at the Andromeda galaxy with the 100-inch telescope. Until then only the outer fringes of the galaxy had been broken up into stars; the central portions had remained a featureless fog. But now Baade had an unusual break. It was wartime and Los Angeles was blacked out. That removed the dim background of distant city light and improved "seeing."

For the first time, photographs were taken that resolved the inner portions of the Andromeda galaxy. Baade could study the very brightest stars of the interior.

It turned out that there were striking differences between the brightest stars of the inner regions and those of the outskirts. The brightest stars in the interior were reddish while those of the outskirts were bluish. That alone accounted for the greater ease with which photographic plates picked up the outer stars, since blue more quickly affects the plates than red does (unless special plates are used). Add to this, the fact that the brightest (bluish) stars on the outskirts were up to a hundred times as bright as the brightest (reddish) stars in the interior.

To Baade, it seemed there were two sets of stars in the Andromeda galaxy with different structures and history. He called the stars of the outskirts Population I, and those of the interior Population II.

Population II is the dominant star-group of the Universe, making up perhaps 98 per cent of the total. They are, by and large, old, moderate-sized stars fairly uniform in characteristics and moving about in dust-free surroundings.

Population I stars are found only in the dust-choked spiral arms of those Galaxies that have spiral arms. On the whole, they are far more scattered in age and structure than the Population II stars, including very young, hot, and luminous stars. (Perhaps Population I stars sweep up the dust through which they pass gradually growing more massive, hotter, and brighter—and shortening their lives, as humans do, by overeating.)

Our own Sun, by the way, happens to be occupying a spiral arm so that the familiar stars of our sky belong to Population I. Our own Sun, fortunately, is an old, quiet, settled star not typical of that turbulent group.

Once the 200-inch telescope was set up on Mount Palomar, Baade continued his investigations of the two populations. There were Cepheid variables in both populations and this brought up an interesting point.

The Cepheids of the Magellanic Clouds (which have no spiral arms) belong to Population II. So do the RR Lyrae variables in the globular clusters. So do the Cepheids of the moderately distant non-globular clusters for which the actual distances were first worked out statistically. In other words all the work done on the size of the Galaxy and the distance of the Magellanic Clouds, as well as on the original establishment of the Cepheid yardstick, were done on Population II Cepheids. So far, so good.

But what about the distance of the outer galaxies? The only stars that could be made out in the outer galaxies such as Andromeda by Hubble and his successors were the extra-large giants of the spiral arms. Those extra-large giants were Population I, and the Cepheids among them were Population I Cepheids. Since Population I is so different from Population II, could one be certain that the Population I Cepheids would fit into a period-luminosity curve which had been worked out from Population II Cepheids only?

Baade began a painstaking comparison of the Population II Cepheids in the globular clusters with the Population I Cepheids in our own neighborhood and in 1952 announced that the latter did *not* fit the Leavitt period-luminosity curve. For any particular period, a Population I Cepheid was between four and five times as luminous as a Population II Cepheid would be. A new period-luminosity curve was drawn for the Population I Cepheids.

Well, then, if the Population I Cepheids of the Andromeda spiral arms were each considerably more than four times brighter than had been thought, then to be as bright as they seemed (the apparent brightness stayed the same of course) they had to be considerably more than twice as far away as had been thought. The flickering yardstick of the Cepheids, which the astronomers had been using to measure the distance of the outer galaxies suddenly turned out to be roughly three times as long as they had thought.

All the nearer galaxies, which had been measured by that yardstick were suddenly pushed a triple distance out into space. The further galaxies whose distances had been estimated by procedures based on the "known" distances of the nearer galaxies, all retreated likewise.

The Universe had again increased in size, and the 200-inch telescope was penetrating, not somewhat less than a billion light-years into space, but a full two billion light-years.

This solved the puzzles of the galaxies. If all the galaxies were about three times the distance that had been thought, they must be larger (in actuality) than had been thought. With all the galaxies suddenly grown up, our own Galaxy is reduced to run-of-the-mill size and is no longer the one out-size member of the family. In fact, the Andromeda galaxy is at least twice the size of ours in terms of numbers of stars contained.

Secondly, the globular clusters around Andromeda, being actually much further away than had been thought, must be more luminous in actual fact than had been thought. Once the greater distance had been allowed for, the globular clusters of Andromeda worked out to be quite comparable to our own in brightness.

Finally, with the galaxies much further spread apart, but with their actual speeds of recession unaffected by the change (the measurement of speed of recession does not depend on the distance of the object being tested) it would take a much longer period for the Universe to have reached its present state from an original compressed hunk of matter. This meant the age of the Universe had to be, at minimum, five or even six billion years. With this figure, geologists were content. They no longer had to consider the Earth to be older than the Universe.

Which was a great relief.

13 *The Sight of Home*

    Now man is struggling toward the Moon but someday, we hope, he will be bouncing among the far stars. Can we imagine that the time may come when some home-sick astronaut will lift his eyes to the strange skies of planets of distant suns in order to locate the tiny speck that is "Sol"? —home, sweet home, across the frigid vastness of space.

A touching picture, but what occurs to me is: How far away can said astronaut be and still make out the sight of home? For that matter, we can make it general and ask: How far away can the inhabitant of any stellar system be and still make out the sight of the star in whose planetary system he was born?

This, of course, depends on how bright the particular star is. I say *is* and not *seems*. From where we sit here on Earth's surface, we see stars of all gradations of brightness. That brightness is partly due to the star's particular luminosity, but is also partly due to the distance it happens to be from us. A star not particularly bright, as stars go, might seem a brilliant specimen to us because it is relatively close; while a star much brighter, but also much more distant, might seem trivial in comparison.

Consider the two stars Alpha Centauri and Capella, for instance. Both are about equally bright in appearance, with magnitudes of about 0.1 and 0.2 respectively. (Remember that the lower the magnitude, the brighter the star, and that each unit decrease of magnitude is equal to a multiplication of 2.52 in brightness.)

However, the two stars are not the same distance from us. Alpha Centauri is the closest of all stars and is only 1.3 parsecs from us. (I am giving all distances in parsecs in this chapter for a reason I will shortly explain. To guide you, a parsec is equal to 3.26 light-years, or to 19,150,000,000,000 miles.) Capella, on the other hand, is 14 parsecs from us, or over 10 times the distance of Alpha Centauri.

Since the intensity of light decreases as the square of the distance, the light of Capella has had a chance to decrease by 10×10 or 100 times more than has the light of Alpha Centauri. Since Capella ends by appearing as bright as Alpha Centauri, it must in reality be 100 times as bright.

If we know a star's distance, we can correct for it. We can calculate what its brightness would be if it were located at some standard distance from us. The distance actually used by astronomers as standard in this connection is 10 parsecs (which is why I am giving all distances as parsecs in this chapter).

Thus the *apparent magnitude* (the actual brightness of a star as we see it) of Alpha Centauri is 0.1 and of Capella is 0.2. The *absolute magnitude*, however (the brightness as it would appear if a star were exactly 10 parsecs away) is 4.8 for Alpha Centauri and –0.6 for Capella.

The Sun, by the way, is just about as bright as Alpha Centauri. Its absolute magnitude is 4.86. Both are average, run-of-the-mill stars.

It is possible to relate absolute magnitude, apparent magnitude, and distance by means of the following simple equation:

$$M = m + 5 - 5 \log D$$

where M is the absolute magnitude of a star, m is the apparent magnitude and D is the distance in parsecs. At the standard distance of 10 parsecs the value of D is 10 and log 10 equals 1. The equation becomes $M = m + 5 - 5$, or $M = m$. The equation at least checks by telling us that at the standard distance of 10 parsecs the apparent magnitude is equal to the absolute magnitude.

But let's use the equation for something more significant. Our astronaut is on a planet of another star and he wants to point out the Sun to the local gentry. He wants to do so with pride, so he would like to have it a first-magnitude star.

The equation will tell us how far away we can be in order that this might be possible. The absolute magnitude of the Sun (M) is 4.86. That can't be changed. We want the apparent magnitude to be 1, so we substitute that for m. We now calculate for D which turns out to be equal to 1.7 parsecs.

Only Alpha Centauri is within 1.7 parsecs of the Sun,

This means from a planet in the Alpha Centauri system only can the Sun be seen as a first-magnitude star, and from no other planetary system in the Universe. Sirius, for instance, is very close to us (less than 3 parsecs away, close enough to be incomparably the brightest star in the sky though only ⅛ as bright as Capella in actuality) and yet even from the Sirian system, the Sun would be seen as only a second-magnitude star.

Well, then, his pride chastened, but homesick nevertheless, our astronaut might abandon first-magnitude pretensions and be willing to settle for any glimpse, however faint, of home.

Since a star of apparent magnitude 6.5 can just barely be made out by a pair of excellent eyes under ideal seeing conditions, let's make *m* equal to 6.5 instead of to 1 and calculate a new value for *D*. Now it comes out as equal to 20 parsecs. The Sun is down to the very limit of naked-eye visibility at a distance of 20 parsecs.

Of course, it is visible for this distance in all directions (assuming that it is not obscured by dust clouds or anything like that) so that it can be seen by naked eye anywhere in a sphere of which the Sun is the center and which has a radius of 20 parsecs. The volume of such a sphere is about 32,000 cubic parsecs.

This sounds like a lot but in the neighborhood of our Sun the density of stars (or multiple stars) is about 4½ per 100 cubic parsecs. Within the visibility sphere of the Sun there are therefore about 1450 stars or multiple-star systems. Since the Galaxy contains about a hundred billion stars, the number of stellar systems from which we can be seen at all, by naked eye, represents an insignificant percentage of those in the Galaxy.

Or put it another way. The Galaxy is about 30,000 parsecs across the full width of its lens-shape. The range of visibility of the Sun is only about ⅟₈₀₀ of this.

Obviously, if we are going to go flitting here and there in the Galaxy, we can just take it for granted that when we lift our tear-filled, homesick eyes to the alien heavens, a sight of home is what we will not get.

Of course, let's not be too sorry for ourselves. There are stars far less luminous than the Sun and therefore far less extensively visible.

The least luminous star known is one which is listed in the books as "Companion of BD + 4°4048" which, for obvious reasons I suggest we call (for purposes of this chapter only) Joe. Now Joe has an absolute magnitude of 19.2. It is only two millionths as bright as the Sun and although it is only about 6 parsecs from us, it is barely visible in a good large telescope.

Using the equation, it turns out that at a distance of 0.03 parsecs, Joe is just barely visible to the naked eye. This means that if Joe were put in the place of the Sun, it would disappear from naked-eye sight at a distance just six times as great as that of the planet Pluto.

It is unlikely that anywhere in the Galaxy there exist two stars this close together, unless, of course, they form part of a multiple-star system. (And Joe *is* part of a multiple-star system, one which includes the star BD + 4°4048, of which it is the "Companion.")

It follows that the existence of a star like Joe would be a complete secret to any race of beings not possessing telescopes and not living on a planet that actually revolves about Joe or about its companion. No man from Joe could ever get a naked-eye sight of home from any planet outside his own multiple system; from any planet at all.

On the other hand, consider stars brighter than the Sun. Sirius, with an absolute magnitude of 1.36 can be made out at a distance of 100 parsecs, while Capella with an absolute magnitude of –0.6 could be seen as far off as 260 parsecs. Sirius could be seen through a volume of space 600 times and Capella through a volume over 2000 times as great as that through which the Sun can be seen.

Nor is Capella the most luminous star by any means. Of all the stars visible to the naked eye, Rigel is about the most luminous. It has an absolute magnitude of –5.8, which makes it over 20,000 times as luminous as the Sun and rather more than 100 times as luminous as even bright Capella.

Rigel can be seen by the naked eye for a distance of 2900 parsecs in any direction, which means over a range of ⅛ the width of the Galaxy. This is rather respectable.

It means that over a large section of the Galaxy we might at least count on identifying our Sun by its spectacular neighbor. We could say to the local Rotarians, "Oh, well you can't see our Sun from here, but it's pretty close to Rigel, that star over there, the one you call Bjfxlpt."

But the record for steady day-in and day-out luminosity is not held by any member of our own Galaxy. There is a star called S Doradus in the Large Magellanic Cloud (which is a kind of satellite galaxy of our own, about 50,000 parsecs away) and S Doradus has an absolute magnitude of –9. It can be seen by naked eye for a distance of 12,500 parsecs. It could be made out all across its own small galaxy and across nearly the full length of our own large one, if it were in our Galaxy.

Of course, no normal star can compete in brightness with a star that explodes. Exploding stars fall into two classes. First there are ordinary novae, which every million years or so blow off a per cent or so of their mass and grow several thousand times brighter (temporarily) when they do so. In between blowoffs they lead fairly normal lives as ordinary stars. Such novae may reach absolute magnitudes of −9, which makes them only as bright as S Doradus is all the time, but then S Doradus is a most unusual star. Certainly the novae are a million times as luminous as are average stars like our Sun.

But then there are supernovae. These are stars that go completely to smash in one big explosion, releasing as much energy in a second as the Sun does in sixty years. Most of their mass is blown off and what is left is converted to a white dwarf. The upper limit of their absolute magnitude reaches anywhere between −14 and −17 so that a large supernova can be 1500 times as luminous as even S Doradus.

If we imagined a good supernova reaching an absolute magnitude of −17, it could be seen by naked eye, at peak brightness, for a distance of 500,000 parsecs. In other words, such a supernova flaring up anywhere in our Galaxy could be seen by naked eye anywhere else in our Galaxy (except where obscured by interstellar dust). It could even be seen in our satellite galaxies, the Large and Small Magellanic Clouds.

However, the distance between our Galaxy and the nearest full-sized neighbor, the Andromeda galaxy, is about 700,000 parsecs. It follows that supernova in other galaxies cannot be seen by naked eye. Any supernova that is visible by naked eye must be located in our own Galaxy, or, at most, in the Magellanic Clouds.

Now astronomers have studied novae which have flared up in our Galaxy. For instance there was a nova in the constellation Hercules in 1934 that rose from telescopic obscurity to the 2nd magnitude (say as bright as the North Star) in a matter of days and stayed near that brightness for three months. In 1942, a nova reached first magnitude (as bright as Arcturus) for a month.

But novae themselves aren't unusual. An average of 20 flare out per year per Galaxy.

Supernovae are different breeds altogether and astronomers would love to get data on them. Unfortunately, they are quite rare. It is estimated that about 3 supernovae appear per galaxy per millennium; that is, one supernova for every 7000 ordinary novae.

Naturally, a supernova can be best studied if it appears in our own Galaxy, and astronomers are waiting for one to appear.

Actually, there is a chance that our Galaxy has had its expected 3 supernovae in the course of the last thousand years. At least there were three very bright novae which have been sighted by naked eye in that interval.

The first of these was sighted in 1054 A.D. by Chinese and Japanese astronomers. From the position in the constellation Taurus, recorded by these Orientals, modern astronomers had a pretty good notion as to where to look for any remnants of the novae. In 1844 the English astronomer, William Parsons, located an odd object in the appropriate place. It was a tiny star barely visible in a good telescope (which eventually turned out to be a white dwarf). Surrounding it was an irregular mass of glowing gas. Because the gas was irregular, with clawlike projections, the object was named the Crab Nebula.

Continued observation over decades showed the gas was expanding. Spectroscopic data revealed the true rate of expansion and that combined with the apparent rate revealed the distance of the Crab Nebula to be about 1600 parsecs. Assuming that the gas had been exploded outward at some time in the past, it was possible to calculate backward to see when that explosion had taken place (from the present position and rate of expansion of the gas). It turns out the explosion took place about 900 years ago. There seems no doubt that the Crab Nebula is what remains of the nova of 1054.

For the nova to be brighter than Venus it must have had a peak apparent magnitude of -5. Substituting that for m in the equation and 1600 for D, the value of M, the absolute magnitude, works out to be just about -16. From this and from the white-dwarf remnant and the gassy explosion, there can be no doubt that the nova of 1054 was a true supernova and one which took place within our Galaxy.

In 1572 a new star appeared in the constellation Cassiopeia. It also outshone Venus and was visible by day. This time it was observed by Europeans. In fact, the last and most famous of all naked-eye astronomers, Tycho Brahe, observed it as a young man and wrote a book about it entitled *De Nova Stella* ("Concerning the New Star") and it is from that title that the word "nova" for new stars comes.

In 1604 still another new star appeared, this time in the constellation Serpens. It was not quite as bright as the nova of 1572 and perhaps only grew as bright as Mars at its brightest (say an apparent magnitude of −2.5). It was observed by another great astronomer, Johann Kepler, who had been Tycho's assistant in the latter's final years.

Now the question is, were the novae of 1572 and 1604 supernovae? Unlike the case of the nova of 1054, no white dwarf, no nebulosity, no anything has been located in the spots reported by Tycho and Kepler. The direct evidence of supernova-hood is missing. Perhaps they were only ordinary novae.

Well, if they were ordinary novae with absolute magnitudes of only −9, then the nova of 1572 must have been about 60 parsecs distant, no more, if it was to surpass Venus in brightness. The nova of 1604 would be 200 parsecs distant. Stars that close could scarcely fail to be seen with modern telescopes, even if they were dim, it seems to me. (Of course, if they ended up as dim as "Joe" they might not be seen, but that level of dimness is most unlikely.)

Most astronomers seem satisfied that the novae of 1572 and 1604 were supernovae in our own Galaxy, and this brings up an irony of astronomical history. Two supernovae appeared in the space of a single generation, the generation just before the invention of the telescope, and not one supernova has appeared in our Galaxy in the nine generations since.

Even a small telescope could have plotted the position of the supernovae more exactly and made it somewhat more likely that the remnant could now be located. Then if the supernovae had appeared after the invention of the spectroscope, things would have been rosier still for happy little astronomers.

As it is, supernovae *have* been observed since Kepler's time, about 50 altogether, but only in other galaxies so that the apparent brightness was so low that little detail could be made out in the spectra.

The brightest and closest supernova to have appeared since 1604 showed up in 1885 in the Andromeda galaxy, our neighbor. It reached an apparent magnitude of 7. (It was not quite visible, you will note, to the naked eye. As I said before, only supernovae in our own Galaxy or in the Magellanic Clouds are visible to the naked eye.) Since the Andromeda galaxy lies at a distance of 700,000 parsecs, the absolute magnitude of the supernova comes out to just a bit brighter than –17. It was about a tenth as bright as the entire galaxy that contained it. Since the Andromeda galaxy is considerably larger than our own, you might say that this one supernova approached the brightness of all the stars of the Milky Way put together—for a while anyway.

(In fact, it was the extraordinary brightness of this star that eventually made astronomers realize that there were novae that were thousands of times brighter than run-of-the-mill novae, and thus the concept of the supernova arose.)

Well now, telescopes and spectroscopes were trained on the supernova of 1885 so that it was better studied than were the much closer ones of 1572 and 1604, but astronomers still weren't living right. Photography had not yet been applied to spectroscopy. If the supernova of 1885 had held on for 20 years more, or if it had been located 20 light-years further from Earth (so that the light would have taken 20 years longer to reach us) its spectrum could have been recorded photographically and studied in detail.

Well, astronomers can only wait! It's even money that sometime during the next century, there will be a supernova blowing its top either in our Galaxy or in the Andromeda galaxy and this time cameras (and heaven knows what else —radio telescopes, too) will be waiting. Provided, of course, the next supernova is not old Sol—the chance of which is, however, virtually nil from what little we know of supernovae.

Still that does bring up a rather grisly situation—the doom of the Earth by nova-formation on the part of the Sun. The Earth would be puffed into gas within minutes of an explosion of the Sun.

Yet is it only the Sun that need concern us? What if a neighboring star exploded?

For instance, suppose it was Alpha Centauri that decided to blow up. If Alpha Centuri became an ordinary nova and reached an absolute magnitude of –9, then its apparent magnitude would be –13.5. It would be two and a half times as bright as the full Moon and a fine spectacle for that portion of Earth's population that lived farther south than Florida and Egypt. (It would be a new tourist attraction and countries like Argentina, the Republic of South Africa, and Australia would clean up for a few months.)

Or suppose Alpha Centauri went supernova and reached an absolute magnitude of −17. (It is impossible for it to do so according to current theories, but let's suppose it anyway.) Its apparent magnitude would then be −21.5, which would make it 4000 times as bright as the full Moon and actually 1/160 as bright as the Sun.

Under such conditions there would be no night for any region of the Earth that had Alpha Centauri in its night sky. You could read newspapers and you would cast a shadow. With Alpha Centauri in the day sky, it would still be a clearly visible and blindingly brilliant point of light and, in the absence of clouds, you would cast a double shadow. In fact, for a couple of months, Earth would be truly a planet of a double sun.

The total amount of energy reaching Earth would (temporarily) be increased by as much as 0.6 per cent. This might have a significant effect on the weather. A large part of the Alpha Centauri radiation would be high-energy and that ought to play hob with the upper atmosphere. In short, although Alpha Centauri as supernova might not endanger life on Earth, it would certainly make things hot for us for a while.

14 Here It Comes; There It Goes

       There's a rumor abroad that I never read any books but my own, but of course that is only a canard. For instance, I have recently read a book called *Towards a Unified Cosmology* by Reginald O. Kapp (Basic Books, 1960) which I enjoyed every bit as much as one of my own.

It presents a view of the Universe, its beginning and its end, so startling, so clearly expressed and so all-but-convincing that I can't resist discussing it. And I must warn you that some of what I say will be mine and not Kapp's and I may not always make it clear which is which.

Kapp considers first the question of the origin of the universe and points out that in general three varieties of outlook are possible.

First, there may have been no origin at all. The matter-energy of the universe may have existed through eternity. This supposition eliminates the nastiness of worrying about creation, perhaps, but it introduces other problems.

For instance, why is the universe in its present active state? Stars are being formed and are converting hydrogen to helium and are eventually being converted to white dwarfs (sometimes going through a nova or supernova stage in the process). If this has been going on through all eternity, why is not all the hydrogen long since converted, all the stars long since exploded or burnt out, all the white dwarfs themselves reduced to black cinders? In short, why is not the universe in a state of maximum entropy?

One way out of the dilemma, which Kapp doesn't mention, but which I once saw suggested is this: The state of maximum entropy is a state of complete randomness. Eventually, by chance movements of the particles in such a universe, a state of partial order is restored; as when by shuffling cards long enough, you manage to get, through pure chance, ten spades in a row. The present active universe may represent such a situation of partially restored order now working its way back to maximum entropy. When that is reached, a period of timeless disorder will ensue until another Universe, perhaps more highly organized than the present one and perhaps less highly organized, is created by chance, and so on.

Another simpler way out of the dilemma is to suppose that the Universe is infinite in extent. It would naturally take an infinite length of time to reduce an infinite universe to maximum entropy. But this piles infinity on infinity and introduces other problems.

The second general hypothesis of origin is that the matter-energy of the universe *was* created all at once at some particular time in the past. This type of theory of origin became popular in the 1920s when the galaxies were found to be hurrying apart at a rate that increased smoothly with their distance.

If we trace matters back into the past, like running a film backward, all the galaxies would suck inward, approach each other, coalesce into one huge gob of matter, the "cosmic egg." It was that egg which is conceived as exploding in the biggest bang in history to start the Universe.

Here there are several sub-possibilities. Either the cosmic egg was created out of nothing and exploded at once, or exploded after an interval of stability. Or else the cosmic egg always existed but happened to explode at one specific time. In any case, special times existed when a creation took place, or an explosion, or both. What would be so special about that time as to bring about so special an event? To answer that one must introduce additional hypotheses. (One such hypothesis, which has been around a long time, is the well-known theological explanation of the Creation.)

Still another sub-possibility is that the Universe first contracts to form a cosmic egg, then expands to some limit, then contracts again, and so on. In such an "oscillating universe" the time of the big bang is merely one extreme of the oscillation and is unusual just because it is an extreme. However, this is also a sub-possibility of the eternal universe theory and involves the problems already mentioned in that connection.

Thus, both varieties of theory as to the beginnings of the Universe involve an original assumption that must then be shored up by additional assumptions, such as an occasionally backward-running entropy, or a periodically contracting universe, or a universe of infinite size.

Now Kapp feels that the necessity of additional hypotheses weakens the original one. He favors a strict application of "Occam's Razor," which is a point of view to the effect that, all things being equal, those explanations of phenomena should be accepted which involve the fewest assumptions. Superfluous assumptions should be shaved away, hence the "razor" part of the phrase, whereas the "Occam" part comes from the fourteenth-century English scholar, William of Occam (or Ockham) who popularized this point of view in a phrase which goes: "Entities must not be unnecessarily multiplied."

Kapp therefore seeks a third type of hypothesis which requires no additional assumptions. This is that creation *does* take place (avoiding the paradoxes of eternal existence) but at no specific time (avoiding the paradoxes of one-shot creation). In other words, at any random point in time and at any random point in space, a particle of matter may be created; not out of energy, mind you, but out of nothing.

Of course you may ask why such a creation should take place, but there is no need to answer that question. The fact of this random creation through space and time is an assumption, but no more an assumption than the hypothesis that matter-energy always existed or that it was all created at once.

Kapp maintains that the assumption of "continuous creation" involves no subsidiary assumptions to justify it and that by the "Principle of Minimum Assumption" (his alternate name for Occam's Razor) it should, at least until further notice, be accepted as the most probable description of the beginnings of the universe.

This continuous-creation theory of the universe has been recently popularized by H. Bondi, Thomas Gold, and, especially, Fred Hoyle, but apparently Kapp got there first. At least he published his suggestions first in 1940, while Hoyle and the others weren't in print on the subject before 1948.

The doctrine of continuous creation raises several interesting questions. First, how quickly is creation going on? At what rate is matter being created? Kapp does not commit himself but quotes an estimate by W. H. McCrea (first published in 1950) to the effect that 500 atoms of hydrogen are being formed per cubic kilometer per year.

If so, the amount of new matter is being formed at a

quite imperceptible rate. To make that clear, consider that the entire volume of the earth is 1.1×10^{12} cubic kilometers, so that in a year the amount of hydrogen that would be created within the planetary body would amount to 5.5×10^{14} atoms. If we allow the earth an existence of four billion years as a solid body (even though the universe as a whole may have no specific time of origin, the earth itself undoubtedly does) and suppose it has occupied the same volume through all that time, the number of hydrogen atoms formed within the earth during its entire existence would be 2.2×10^{24}.

That's over two trillion trillion atoms, which may sound like a lot, but only comes to about 3.6 grams or less than $\frac{1}{7}$ of an ounce. I think you'll agree that this addition to the earth's mass would go unnoticed by even our best instruments working through earth's entire history.

However, the *total* amount of matter so created is enormous. Consider a sphere of space with a radius of one billion light-years (a volume that is certainly smaller than the observable universe). Its volume is about 4×10^{66} cubic kilometers and in one year, the number of hydrogen atoms formed throughout that volume is equal to 2×10^{69}. This number of hydrogen atoms can be used to form something more than a trillion suns like ours or about ten galaxies as large as our own. A process which creates enough matter for ten galaxies each year is not to be shrugged off.

But *what* is being created? The universe is 90 per cent hydrogen and most of what remains is the helium originally formed in the center of stars as a result of thermonuclear reactions. It seems reasonable that if the stars weren't working hard, the universe would consist only of hydrogen, the simplest of all atoms. Does it not seem reasonable that it is hydrogen (as McCrea implies) that is being formed?

The trouble is that the hydrogen atom is itself composite, containing one proton and one electron. Are they created separately? Does that mean there are two kinds of creation that keep in step so that just as many protons are formed as electrons?

Kapp shrugs off the issue by refusing to pinpoint the exact nature of the matter being created. I myself will run the risk of suggesting that it may be the neutron. A neutron, in the course of nature, quickly decays to produce a proton and an electron (and an anti-neutrino, which we will ignore). The protons and electrons formed from the neutrons, just about as fast as the latter are created, will associate to form hydrogen atoms.

But why should neutrons be created and not anti-neutrons? There seems to me no reason for supposing that there is a greater probability of the creation of any particle than of the corresponding anti-particle.

Whatever the mechanism of creation; whether hydrogen atoms are created to begin with, or neutrons, or some unknown and still more fundamental particle, it seems to me that on the basis of pure chance, matter and anti-matter should be formed in equal quantities. What's more, they should be formed randomly mixed throughout space and time. Matter and anti-matter should then interact and produce a universe consisting of pure energy. Nothing in Kapp's book satisfies me as a way out of this dilemma.

But let's put that to one side and proceed.

Kapp goes on to consider the ultimate end of the universe. Again, he reduces all speculations to three possible varieties of assumption: one, that the mass-energy of the Universe will exist through all future eternity; two, that it will all come to an end at once at some specific time; three, that individual particles will cease to exist at random at any time and in any place.

Using the same sort of reasoning as before, he plumps for the third possibility and again, I, for one, find the reasoning all but irresistible and feel the strong urge to go along with him.

So Kapp, having anticipated the continuous-creation boys, goes beyond them by suggesting the existence of continuous extinction as well. The two together he calls "The Hypothesis of the Symmetrical Impermanence of Matter"; i.e. that matter is impermanent in its past history and in its future history, and in the same statistical manner.

For any given particle of matter, then, it is a case of "here it comes and there it goes."

If matter is being created and extinguished constantly, there is the possibility that both processes are proceeding at equal rates so that the total matter-energy of the universe remains constant, even though the identity of individual particles would be constantly changing. (We would then be living in a "steady-state universe.")

This would seem unlikely, at least at the present stage of the universe's existence. The creation of a particle of matter would create an increment of space as well, while the extinction of a particle would extinguish an increment of space. (Space, in this view, is not merely an empty container into which matter is piled, but is an integral part of matter, just as mass is, coming with matter and going with it.)

Since the universe is observed to be expanding, this would seem to require that those processes that create space preponderate over the processes that extinguish it. McCrea apparently determined the rate at which matter is created by computing the amount of space that had to be added to the universe to account for its observed rate of expansion.

If Kapp's suggestion of continuous extinction is accepted, then the hydrogen atoms being formed (according to McCrea) are not the total being formed. They merely represent the excess of creations over extinctions.

However, just as there is the question of particle/antiparticle balance which seems to me to be a weakness in the hypothesis of continuous creations, so there is a question of another sort which bothers me with respect to continuous extinction.

Kapp himself points out that it is unlikely that a single particle of a complex nucleus will be extinguished alone. That could easily render what remains of the nucleus radioactive. If (to use an example of my own) one of the neutrons of the argon-40 nucleus were to vanish suddenly, the strongly radioactive argon-39 would be formed. If, instead, one of the protons were to disappear, the even more strongly radioactive chlorine-39 should appear.

In that case, the extinction of matter in a pure sample of argon-40 should be detectable, even if it proceeds at an excessively slow rate, through the appearance of radioactivity. However, argon-40 is not detectably radioactive.

Kapp therefore concludes that the smallest particle that can be involved in the process of extinction is the atomic nucleus which must go poof as a whole. If this were so, then continuous extinction could only be detected through disappearing mass, a much more difficult phenomenon to measure at micro-micro-levels than is appearing radiation.

But then this means that the two hundred odd protons and neutrons (plus mesons and who knows what else) in complex nuclei such as those of mercury or uranium must all go at once and together.

Why?

The particles come in singly, so why go out in a group? What keeps them so neatly in step? Does their close association in the nucleus make them all one particle in certain aspects? Do we not require additional assumptions here, and does this not, in view of Occam's Razor, weaken the hypothesis of continuous extinction?

Now although the universe may see an over-all excess of creations over extinctions, this is not necessarily true for a specific small portion of the universe. Creations take place anywhere in space and time randomly, so that a cubic kilometer which is virtually empty of matter (as in intergalactic space) and a cubic kilometer which is virtually full of it (as at the center of a planet) witness creations at equal rates. Creation, in other words, is a function of volume only.

Extinctions, on the other hand, depend on particles already existing. In those regions of space where particles are almost non-existent, there are virtually no extinctions because there is nothing to extinguish. In other regions, where particles exist cheek by jowl, there are comparatively many extinctions. In short, extinctions are a function of mass only.

Therefore wherever much mass is compressed into comparatively small volume, as in a planet, extinctions overwhelm creations and the net effect is a local shrinkage of the universe. Where a minute mass is distributed over vast volumes, creations overwhelm extinctions and there is a local expansion of the universe. On the whole, as I've said, the expansion preponderates over the contraction.

Now consider two galaxies which are neighbors. Between them is a vast region of space, virtually empty, in which creations of particles are proceeding at a considerably greater rate than are extinctions, so that space is expanding and the galaxies are receding from each other. (The recession is not caused by the motion of the galaxies but by the piling up of space between, if you can picture the distinction.)

Although space comes into existence with matter and is part of it, the matter, once created, can move about in space under the influence of gravitational attraction, crowding into some portions and leaving others emptier than ever. In this case, the particles formed between the galaxies move slowly toward whichever galaxy has the stronger gravitational pull at that point in space.

However, halfway between the galaxies (assuming them to be equal in mass) there is a kind of gravitational plateau where the particles move in either direction with such excessive slowness as to be considered virtually motionless.

The further apart the galaxies move, the vaster is this intermediate region in which the created particles are virtually motionless. The result is that they begin to accumulate and after a while develop a gravitational field of their own strong enough to draw them together against the pull of the distant galaxies. The compression further strengthens the new gravitational field and the new mass now begins to attract particles on either side that otherwise would have fallen into the old galaxies.

In short, a new galaxy is formed.

Kapp calculates that the universe is expanding at such a rate that a new galaxy forms between two old neighbors after those neighbors have been mutually receding for a little over three and a half billion years. The space between the new galaxy and each of its neighbors continues to increase and after another three and a half billion years, still newer galaxies form between it and its neighbors on all sides.

In the volume occupied by any piece of dense matter, such as a gram weight or a planet, the number of extinctions far exceeds the number of creations and the mass of matter constantly decreases. Since extinctions take place on a purely random basis, as radioactive disintegrations do, the "half-life" concept holds. That is, after a fixed period of time, a given mass will have shrunk to half its original value. After the lapse of another fixed period, the remainder will have shrunk to half what it was and so on.

Kapp deduces by several lines of reasoning that the half-life of matter is roughly 800,000,000 years, which is an astonishingly small value. It means that some 300,000,000,000 atomic nuclei are undergoing extinction every second in your body. This isn't as bad as it sounds, of course, since the mass of that number of nuclei is less than a thirty-trillionth of an ounce and is made up without detectable effort.

However, the consequences in geology and astronomy are more drastic. Kapp suggests that a large body such as a star may make up its shrinkage by the collection of interstellar matter through gravitational attraction. For that reason a star may be undergoing only a very small net shrinkage or may even, if massive enough, be growing.

A smaller body in a star's shadow, so to speak, has little or no chance to collect matter, since the star, with its larger gravitational field, hogs the collection. The smaller body will decrease in mass, therefore, and the smaller it is, the more closely will its rate of mass decrease approach the half-life of matter. If the body is sizable, its gravity will keep it compact, causing it to shrink in volume as it decreases in mass.

In fact, Kapp works up a theory of the formation of the Solar System as the result of the shrinkage of such a small companion of our Sun, and maintains that what is left of that small companion is what is now called the planet Jupiter.

The mass of Jupiter at present is just a trifle under a thousandth that of the Sun; 0.00095, to be precise. Suppose we assume that Jupiter has been shrinking at a rate corresponding to Kapp's estimated half-life of matter and that the Sun has been maintaining a constant mass. If that is so, then about eight billion years ago, Jupiter would have been just as massive as the Sun. Since Kapp's theory of the formation of the Solar System postulates a companion markedly less massive than the Sun to begin with, the Solar System must be markedly younger than eight billion years.

And so it is, in all likelihood. The most popular estimate of the age of the Solar System is five billion years, and that long ago Jupiter would have been 0.0788 (about one-thirteenth) the mass of the Sun. This is a reasonable mass for a small star.

The planets, including the earth itself, must also be shrinking. From this point of view, the earth would have shrunk considerably during geologic times.

If life began two billion years ago, it began on an earth that was 5.6 times as massive as it is today and had a diameter of about 14,000 miles. Six hundred million years ago, at a time when the earliest fossils were formed, the earth was still 1.7 times as massive as it is today and had a diameter of 9500 miles. A hundred fifty million years ago, when the dinosaurs flourished, the earth was 1.2 times as massive as it is today and had a diameter of 8500 miles.

And, of course, this shrinkage continues. In about two and a half billion years, the earth will be no more massive than present-day Mars; most of its atmosphere would be gone and most of its ocean. A dreary picture.

Of all Kapp's suggestions, I find the notion of the shrinking earth most difficult to swallow. What I would like to see is some observation that would present tangible evidence for or against such a shrinkage.

The most obvious method would be to measure the strength of the earth's gravitational field and note if it decreases slowly with time. Unfortunately, this decrease would be excessively slow. The acceleration of a falling body under standardized conditions is now 980.665 centimeters per second per second. If Kapp is correct, it will decrease to

980.663 by 2250 A.D. Three centuries is a long time to wait for a decline of one part in half a million.

However, I have thought (and I absolve Kapp of responsibility for this idea) of a way in which the question might be settled right now.

If an animal doubles in dimensions, its mass (which would depend on its volume) would increase as the cube of the dimensional increase, or eightfold. On the other hand, the strength of supporting structures (such as the bones of the limbs) would increase only as the cross-sectional area, or fourfold.

For this reason, a massive animal must have thicker legs, even in proportion to its size, than a small animal. The legs of an elephant are thicker in proportion to its body size than are the legs of a horse, which are in turn thicker than those of a mouse, which are in turn thicker than those of a mosquito.

If an animal the size and shape of a horse lived on a world with a greater gravitational attraction than earth's it would have to have markedly thicker legs than it does now. If it lived on a world with a smaller attraction, it would have thinner legs.

Now at the time the dinosaurs were in their prime, the earth was 1.2 times as massive as it is today, according to Kapp's theory. The fossilized bones we now have would also have been 1.2 times as massive when they were living. The gravitational attraction of earth upon dinosaur would therefore be 1.2 × 1.2 or just about 1½ times as great as we would expect it to be from today's size of the planet and the fossil. A fossil which, under present-day conditions, we would estimate to represent a dinosaur that weighed 40 tons, would really be representing one that weighed 60 tons. (In the case of the first land creatures, such as the armored amphibians of three hundred million years ago, the discrepancy would be even greater.)

Now underground pressures ought to keep the fossils compact and the shrinkage of the fossils ought to be perfectly even as mass disappears, maintaining all bone or shell proportions as they originally were. Would it be possible for a paleontologist, then, to tell from these proportions whether the bones were more suitable to a 60-ton mass than to a 40-ton mass, or vice versa? It seems to me that it should be, but is there a paleontologist in the house?

PART. IV. / THE HUMAN MIND

15 *Those Crazy Ideas*

Time and time again I have been asked (and I'm sure others who have, in their time, written science fiction have been asked too): "Where do you get your crazy ideas?"

Over the years, my answers have sunk from flattered confusion to a shrug and a feeble smile. Actually, I don't really know, and the lack of knowledge doesn't really worry me, either, as long as the ideas keep coming.

But then some time ago, a consultant firm in Boston, engaged in a sophisticated space-age project for the government, got in touch with me.

What they needed, it seemed, to bring their project to a successful conclusion were novel suggestions, startling new principles, conceptual breakthroughs. To put it into the nutshell of a well-turned phrase, they needed "crazy ideas."

Unfortunately, they didn't know how to go about getting crazy ideas, but some among them had read my science fiction, so they looked me up in the phone book and called me to ask (in essence), "Dr. Asimov, where do you get your crazy ideas?"

Alas, I still didn't know, but as speculation is my profession, I am perfectly willing to think about the matter and share my thoughts with you.

The question before the house, then, is: How does one go about creating or inventing or dreaming up or stumbling over a new and revolutionary scientific principle?

For instance—to take a deliberately chosen example—how did Darwin come to think of evolution?

To begin with, in 1831, when Charles Darwin was twenty-two, he joined the crew of a ship called the *Beagle*. This ship was making a five-year voyage about the world to explore various coast lines and to increase man's geographical knowledge. Darwin went along as ship's naturalist, to study the forms of life in far-off places.

This he did extensively and well, and upon the return of the *Beagle* Darwin wrote a book about his experiences (published in 1840) which made him famous. In the course of this voyage, numerous observations led him to the conclusion that species of living creatures changed and developed slowly with time; that new species descended from old. This, in itself was not a new idea. Ancient Greeks had had glimmerings of evolutionary notions. Many scientists before Darwin, including Darwin's own grandfather, had theories of evolution.

The trouble, however, was that no scientist could evolve an explanation for the *why* of evolution. A French naturalist, Jean Baptiste de Lamarck, had suggested in the early 1800s that it came about by a kind of conscious effort or inner drive. A tree-grazing animal, attempting to reach leaves, stretched its neck over the years and transmitted a longer neck to its descendants. The process was repeated with each generation until a giraffe in full glory was formed.

The only trouble was that acquired characteristics are not inherited and this was easily proved. The Lamarckian explanation did not carry conviction.

Charles Darwin, however, had nothing better to suggest after several years of thinking about the problem.

But in 1798, eleven years before Darwin's birth, an English clergyman named Thomas Robert Malthus, had written a book entitled *An Essay on the Principle of Population*. In this book Malthus suggested that the human population always increased faster than the food supply and that the population had to be cut down by either starvation, disease, or war; that these evils were therefore unavoidable.

In 1838 Darwin, still puzzling over the problem of the development of species, read Malthus's book. It is hackneyed to say "in a flash" but that, apparently, is how it happened. In a flash, it was clear to Darwin. Not only human beings increased faster than the food supply; all species of living things did. In every case, the surplus population had to be cut down by starvation, by predators, or by disease. Now no two members of any species are exactly alike; each has slight individual variations from the norm. Accepting this fact, which part of the population was cut down?

Why—and this was Darwin's breakthrough—those members of the species who were less efficient in the race for food, less adept at fighting off or escaping from predators, less equipped to resist disease, went down.

The survivors, generation after generation, were better adapted, on the average, to their environment. The slow changes toward a better fit with the environment accumulated until a new (and more adapted) species had replaced the old. Darwin thus postulated the reason for evolution as being the action of *natural selection*. In fact, the full title of his book is *On the Origin of Species by Means of Natural Selection, or the Preservation of Favoured Races in the Struggle for Life*. We just call it *The Origin of Species* and miss the full flavor of what it was he did.

It was in 1838 that Darwin received this flash and in 1844 that he began writing his book, but he worked on for fourteen years gathering evidence to back up his thesis. He was a methodical perfectionist and no amount of evidence seemed to satisfy him. He always wanted more. His friends read his preliminary manuscripts and urged him to publish. In particular, Charles Lyell (whose book *Principles of Geology*, published in 1830–1833, first convinced scientists of the great age of the earth and thus first showed there was *time* for the slow process of evolution to take place) warned Darwin that someone would beat him to the punch.

While Darwin was working, another and younger English naturalist, Alfred Russel Wallace, was traveling in distant lands. He too found copious evidence to show that evolution took place and he too wanted to find a reason. He did not know that Darwin had already solved the problem.

He spent three years puzzling, and then in 1858, he too came across Malthus's book and read it. I am embarrassed to have to become hackneyed again, but in a flash he saw the answer. Unlike Darwin, however, he did not settle down to fourteen years of gathering and arranging evidence.

Instead, he grabbed pen and paper and at once wrote up his theory. He finished this in two days.

Naturally, he didn't want to rush into print without having his notions checked by competent colleagues, so he decided to send it to some well-known naturalist. To whom? Why, to Charles Darwin. To whom else?

I have often tried to picture Darwin's feeling as he read Wallace's essay which, he afterward stated, expressed matters in almost his own words. He wrote to Lyell that he had been forestalled "with a vengeance."

Darwin might easily have retained full credit. He was well-known and there were many witnesses to the fact that he had been working on his project for a decade and a half. Darwin, however, was a man of the highest integrity. He made no attempt to suppress Wallace. On the contrary, he passed on the essay to others and arranged to have it published along with a similar essay of his own. The year after, Darwin published his book.

Now the reason I chose this case was that here we have two men making one of the greatest discoveries in the history of science independently and simultaneously and under precisely the same stimulus. Does that mean *anyone* could have worked out the theory of natural selection if he had but made a sea voyage and combined that with reading Malthus?

Well, let's see. Here's where the speculation starts.

To begin with, both Darwin and Wallace were thoroughly grounded in natural history. Each had accumulated a vast collection of facts in the field in which they were to make their breakthrough. Surely this is significant.

Now every man in his lifetime collects facts, individual pieces of data, items of information. Let's call these "bits" (as they do, I think, in information theory). The "bits" can be of all varieties: personal memories, girls' phone numbers, baseball players' batting averages, yesterday's weather, the atomic weights of the chemical elements.

Naturally, different men gather different numbers of different varieties of "bits." A person who has collected a larger number than usual of those varieties that are held to be particularly difficult to obtain—say, those involving the sciences and the liberal arts—is considered "educated."

There are two broad ways in which the "bits" can be accumulated. The more common way, nowadays, is to find people who already possess many "bits" and have them transfer those "bits" to your mind in good order and in predigested fashion. Our schools specialize in this transfer of "bits" and those of us who take advantage of them receive a "formal education."

The less common way is to collect "bits" with a minimum amount of live help. They can be obtained from books or out of personal experience. In that case you are "self-educated." (It often happens that "self-educated" is confused with "uneducated." This is an error to be avoided.)

In actual practice, scientific breakthroughs have been initiated by those who were formally educated, as for instance by Nicolaus Copernicus, and by those who were self-educated, as for instance by Michael Faraday.

To be sure, the structure of science has grown more complex over the years and the absorption of the necessary number of "bits" has become more and more difficult without the guidance of someone who has already absorbed them. The self-educated genius is therefore becoming rarer, though he has still not vanished.

However, without drawing any distinction according to the manner in which "bits" have been accumulated, let's set up the first criterion for scientific creativity:

1) The creative person must possess as many "bits" of information as possible; i.e. he must be educated.

Of course, the accumulation of "bits" is not enough in itself. We have probably all met people who are intensely educated, but who manage to be abysmally stupid, nevertheless. They have the "bits," but the "bits" just lie there.

But what is there one can do with "bits"?

Well, one can combine them into groups of two or more. Everyone does that; it is the principle of the string on the finger. You tell yourself to remember *a* (to buy bread) when you observe *b* (the string). You enforce a combination that will not let you forget *a* because *b* is so noticeable.

That, of course, is a conscious and artificial combination of "bits." It is my feeling that every mind is, more or less unconsciously, continually making all sorts of combinations and permutations of "bits," probably at random.

Some minds do this with greater facility than others; some minds have greater capacity for dredging the combinations out of the unconscious and becoming consciously aware of them. This results in "new ideas," in "novel outlooks."

The ability to combine "bits" with facility and to grow consciously aware of the new combinations is, I would like to suggest, the measure of what we call "intelligence." In this view, it is quite possible to be educated and yet not intelligent.

Obviously, the creative scientist must not only have his "bits" on hand but he must be able to combine them readily and more or less consciously. Darwin not only observed data, he also made deductions—clever and far-reaching deductions —from what he observed. That is, he combined the "bits" in interesting ways and drew important conclusions.

So the second criterion of creativity is:

2) The creative person must be able to combine "bits" with facility and recognize the combinations he has formed; i.e. he must be intelligent.

Even forming and recognizing new combinations is insufficient in itself. Some combinations are important and some are trivial. How do you tell which are which? There is no question but that a person who cannot tell them apart must labor under a terrible disadvantage. As he plods after each possible new idea, he loses time and his life passes uselessly.

There is also no question but that there are people who somehow have the gift of seeing the consequences "in a flash" as Darwin and Wallace did; of feeling what the end must be without consciously going through every step of the reasoning. This, I suggest, is the measure of what we call "intuition."

Intuition plays more of a role in some branches of scientific knowledge than others. Mathematics, for instance, is a deductive science in which, once certain basic principles are learned, a large number of items of information become "obvious" as merely consequences of those principles. Most of us, to be sure, lack the intuitive powers to see the "obvious."

To the truly intuitive mind, however, the combination of the few necessary "bits" is at once extraordinarily rich in consequences. Without too much trouble they see them all, including some that have not been seen by their predecessors.[1]

It is perhaps for this reason that mathematics and mathematical physics have seen repeated cases of first-rank breakthroughs by youngsters. Evariste Galois evolved group theory at twenty-one. Isaac Newton worked out calculus at twenty-three. Albert Einstein presented the theory of relativity at twenty-six, and so on.

In those branches of science which are more inductive and require larger numbers of "bits" to begin with, the average age of the scientists at the time of the breakthrough is greater. Darwin was twenty-nine at the time of his flash, Wallace was thirty-five.

But in any science, however inductive, intuition is necessary for creativity. So:

3) The creative person must be able to see, with as little delay as possible, the consequences of the new combinations of "bits" which he has formed; i.e. he must be intuitive.

But now let's look at this business of combining "bits" in a little more detail. "Bits" are at varying distances from each

---

[1] The Swiss mathematician, Leonhard Euler, said that to the true mathematician, it is at once obvious that $e\pi^1 = -1$.

other. The more closely related two "bits" are, the more apt one is to be reminded of one by the other and to make the combination. Consequently, a new idea that arises from such a combination is made quickly. It is a "natural consequence" of an older idea, a "corollary." It "obviously follows."

The combination of less related "bits" results in a more startling idea; if for no other reason than that it takes longer for such a combination to be made, so that the new idea is therefore less "obvious." For a scientific breakthrough of the first rank, there must be a combination of "bits" so widely spaced that the random chance of the combination being made is small indeed. (Otherwise, it will be made quickly and be considered but a corollary of some previous idea which will then be considered the "breakthrough.")

But then, it can easily happen that two "bits" sufficiently widely spaced to make a breakthrough by their combination are not present in the same mind. Neither Darwin nor Wallace, for all their education, intelligence, and intuition, possessed the key "bits" necessary to work out the theory of evolution by natural selection. Those "bits" were lying in Malthus's book, and both Darwin and Wallace had to find them there.

To do this, however, they had to read, understand, and appreciate the book. In short, they had to be ready to incorporate other people's "bits" and treat them with all the ease with which they treated their own.

It would hamper creativity in other words, to emphasize intensity of education at the expense of broadness. It is bad enough to limit the nature of the "bits" to the point where the necessary two would not be in the same mind. It would be fatal to mold a mind to the point where it was incapable of accepting "foreign bits."

I think we ought to revise the first criterion of creativity, then, to read:

1) The creative person must possess as many "bits" as possible, falling into as wide a variety of types as possible; i.e. he must be broadly educated.

As the total amount of "bits" to be accumulated increases with the advance of science, it is becoming more and more difficult to gather enough "bits" in a wide enough area. Therefore, the practice of "brain-busting" is coming into popularity; the notion of collecting thinkers into groups and hoping that they will cross-fertilize one another into startling new breakthroughs.

Under what circumstances could this conceivably work?

(After all, anything that will stimulate creativity is of first importance to humanity.)

Well, to begin with, a group of people will have more "bits" on hand than any member of the group singly since each man is likely to have some "bits" the others do not possess.

However, the increase in "bits" is not in direct proportion to the number of men, because there is bound to be considerable overlapping. As the group increases, the smaller and smaller addition of completely new "bits" introduced by each additional member is quickly outweighed by the added tensions involved in greater numbers; the longer wait to speak, the greater likelihood of being interrupted, and so on. It is my (intuitive) guess that five is as large a number as one can stand in such a conference.

Now of the three criteria mentioned so far, I feel (intuitively) that intuition is the least common. It is more likely that none of the group will be intuitive than that none will be intelligent or none educated. If no individual in the group is intuitive, the group as a whole will not be intuitive. You cannot add non-intuition and form intuition.

If one of the group is intuitive, he is almost certain to be intelligent and educated as well, or he would not have been asked to join the group in the first place. In short, for a brain-busting group to be creative, it must be quite small and it must possess at least one creative individual. But in that case, does that one individual need the group? Well, I'll get back to that later.

Why did Darwin work fourteen years gathering evidence for a theory he himself must have been convinced was correct from the beginning? Why did Wallace send his manuscript to Darwin first instead of offering it for publication at once?

To me it seems that they must have realized that any new idea is met by resistance from the general population who, after all, are not creative. The more radical the new idea, the greater the dislike and distrust it arouses. The dislike and distrust aroused by a first-class breakthrough are so great that the author must be prepared for unpleasant consequences (sometimes for expulsion from the respect of the scientific community; sometimes, in some societies, for death).

Darwin was trying to gather enough evidence to protect himself by convincing others through a sheer flood of reasoning. Wallace wanted to have Darwin on his side before proceeding.

It takes courage to announce the results of your creativity. The greater the creativity, the greater the necessary courage in much more than direct proportion. After all, consider that the more profound the breakthrough, the more solidified the previous opinions; the more "against reason" the new discovery seems; the more against cherished authority.

Usually a man who possesses enough courage to be a scientific genius seems odd. After all, a man who has sufficient courage or irreverence to fly in the face of reason or authority *must* be odd, if you define "odd" as "being not like most people." And if he is courageous and irreverent in such a colossally big thing, he will certainly be courageous and irreverent in many small things so that being odd in one way, he is apt to be odd in others. In short, he will seem to the non-creative, conforming people about him to be a "crackpot."

So we have the fourth criterion:

4) The creative person must possess courage (and to the general public may, in consequence, seem a crackpot).

As it happens, it is the crackpottery that is most often most noticeable about the creative individual. The eccentric and absent-minded professor is a stock character in fiction; and the phrase "mad scientist" is almost a cliché.

(And be it noted that I am never asked where I get my interesting or effective or clever or fascinating ideas. I am invariably asked where I get my *crazy* ideas.)

Of course, it does not follow that because the creative individual is usually a crackpot, that any crackpot is automatically an unrecognized genius. The chances are low indeed, and failure to recognize that the proposition can not be so reversed is the cause of a great deal of trouble.

Then, since I believe that combinations of "bits" take place quite at random in the unconscious mind, it follows that it is quite possible that a person may possess all four of the criteria I have mentioned in superabundance and yet may never happen to make the necessary combination. After all, suppose Darwin had never read Malthus. Would he ever have thought of natural selection? What made him pick up the copy? What if someone had come in at the crucial time and interrupted him?

So there is a fifth criterion which I am at a loss to phrase in any other way than this:

5) A creative person must be lucky.

To summarize:

A creative person must be 1) broadly educated, 2) intelligent, 3) intuitive, 4) courageous, and 5) lucky.

How, then, does one go about encouraging scientific creativity? For now, more than ever before in man's history, we must; and the need will grow constantly in the future.

Only, it seems to me, by increasing the incidence of the various criteria among the general population.

Of the five criteria, number 5 (luck) is out of our hands. We can only hope; although we must also remember Louis Pasteur's famous statement that "Luck favors the prepared mind." Presumably, if we have enough of the four other criteria, we shall find enough of number five as well.

Criterion 1 (broad education) is in the hands of our school system. Many educators are working hard to find ways of increasing the quality of education among the public. They should be encouraged to continue doing so.

Criterion 2 (intelligence) and 3 (intuition) are inborn and their incidence cannot be increased in the ordinary way. However, they can be more efficiently recognized and utilized. I would like to see methods devised for spotting the intelligent and intuitive (particularly the latter) early in life and treating them with special care. This, too, educators are concerned with.

To me, though, it seems that it is criterion 4 (courage) that receives the least concern, and it is just the one we may most easily be able to handle. Perhaps it is difficult to make a person more courageous than he is, but that is not necessary. It would be equally effective to make it sufficient to be less courageous; to adopt an attitude that creativity is a permissible activity.

Does this mean changing society or changing human nature? I don't think so. I think there are ways of achieving the end that do not involve massive change of anything, and it is here that brain-busting has its greatest chance of significance.

Suppose we have a group of five that includes one creative individual. Let's ask again what that individual can receive from the non-creative four?

The answer to me, seems to be just this: Permission!

They must permit him to create. They must tell him to go ahead and be a crackpot.[2]

How is this permission to be granted? Can four essentially non-creative people find it within themselves to grant such permission? Can the one creative person find it within himself to accept it?

I don't know. Here, it seems to me, is where we need experimentation and perhaps a kind of creative breakthrough about creativity. Once we learn enough about the whole matter, who knows—I may even find out where I get those crazy ideas.

[2] Always with the provision of course, that the crackpot creation that results survives the test of hard inspection. Though many of the products of genius seem crackpot at first, very few of the creations that seem crackpot turn out, after all, to be products of genius. I shall go into that aspect of the matter in the next chapter.

16 *My Built-in Doubter*

> *Once I delivered myself of an oration before*
a small but select audience of non-scientists on the topic
of "What Is Science?" speaking seriously and, I hope,
intelligently.

Having completed the talk, there came the question period,
and, bless my heart, I wasn't disappointed. A charming young
lady up front waved a pretty little hand at me and asked,
not a serious question on the nature of science, but: "Dr.
Asimov, do you believe in flying saucers?"

With a fixed smile on my face, I proceeded to give the
answer I have carefully given after every lecture I have delivered. I said, "No, miss, I do not, and I think anyone who
does is a crackpot!"*

And oh, the surprise on her face!

It is taken for granted by everyone, it seems to me, that
because I sometimes write science fiction, I believe in flying
saucers, in Atlantis, in clairvoyance and levitation, in the
prophecies of the Great Pyramid, in astrology, in Fort's theories, and in the suggestion that Bacon wrote Shakespeare.

No one would ever think that someone who writes fan-
tasies for pre-school children really thinks that rabbits can
talk, or that a writer of hard-boiled detective stories really
thinks a man can down two quarts of whiskey in five minutes,
then make love to two girls in the next five, or that a writer

---

* Since this article first appeared, I have received strong ob-
jections to the use of the word from flying-saucer fanciers. Let me
stress that it is not intended to apply to those who suspect that we
are not yet aware of the significance of all atmospheric phenomena and that "unidentified flying objects" are therefore a reasonable object of scientific study. However, what my questioner
and I meant was "flying saucer" in the sense of a spaceship carrying little green men from Venus—or the equivalent. Those who
believe this, I repeat, are, in my opinion, crackpots. I. A.

for the ladies' magazines really thinks that virtue always triumphs and that the secretary always marries the handsome boss—but a science-fiction writer apparently *must* believe in flying saucers.

Well, I do not.

To be sure, I wrote a story once about flying saucers in which I explained their existence very logically. I also wrote a story once in which levitation played a part.

If I can buddy up to such notions long enough to write sober, reasonable stories about them, why, then, do I reject them so definitely in real life?

I can explain by way of a story. A good friend of mine once spent quite a long time trying to persuade me of the truth and validity of what I considered a piece of pseudo-science and bad pseudo-science at that. I sat there listening quite stonily, and none of the cited evidence and instances and proofs had the slightest effect on me.

Finally the gentleman said to me, with considerable annoyance, "Damn it, Isaac, the trouble with you is that you have a built-in doubter."

To which the only answer I could see my way to making was a heartfelt, "Thank God."

If a scientist has one piece of temperamental equipment that is essential to his job, it is that of a built-in doubter. Before he does anything else, he must doubt. He must doubt what others tell him and what he reads in reference books, and, *most of all,* what his own experiments show him and what his own reasoning tells him.

Such doubt must, of course, exist in varying degrees. It is impossible, impractical, and useless to be a maximal doubter at all times. One cannot (and would not want to) check personally every figure or observation given in a handbook or monograph, before one uses it and then proceed to check it and recheck it until one dies. *But,* if any trouble arises and nothing else seems wrong, one must be prepared to say to one's self, "Well, now, I wonder if the data I got out of the 'Real Guaranteed Authoritative Very Scientific Handbook' might not be a misprint."

To doubt intelligently requires, therefore, a rough appraisal of the authoritativeness of a source. It also requires a rough estimate of the nature of the statement. If you were to tell me that you had a bottle containing one pound of pure titanium oxide, I would say, "Good," and ask to borrow some if I needed it. Nor would I test it. I would accept its purity on your say-so (until further notice, anyway).

If you were to tell me that you had a bottle containing one pound of pure thulium oxide, I would say with considerable astonishment, "You have? Where?" Then if I had use for the stuff, I would want to run some tests on it and even run it through an ion-exchange column before I could bring myself to use it.

And if you told me that you had a bottle containing one pound of pure americium oxide, I would say, "You're crazy," and walk away. I'm sorry, but my time is reasonably valuable, and I do not consider that statement to have enough chance of validity even to warrant my stepping into the next room to look at the bottle.

What I am trying to say is that doubting is far more important to the advance of science than believing is and that, moreover, doubting is a serious business that requires extensive training to be handled properly. People without training in a particular field do not know what to doubt and what not to doubt; or, to put it conversely, what to believe and what not to believe. I am very sorry to be undemocratic, but one man's opinion is not necessarily as good as the next man's.

To be sure, I feel uneasy about seeming to kowtow to authority in this fashion. After all, you all know of instances where authority was wrong, dead wrong. Look at Columbus, you will say. Look at Galileo.

I know about them, and about others, too. As a dabbler in the history of science, I can give you horrible examples you may never have heard of. I can cite the case of the German scientist, Rudolf Virchow, who, in the mid-nineteenth century was responsible for important advances in anthropology and practically founded the science of pathology. He was the first man to engage in cancer research on a scientific basis. However, he was dead set against the germ theory of disease when that was advanced by Pasteur. So were many others, but one by one the opponents abandoned doubt as evidence multiplied. Not Virchow, however. Rather than be forced to admit he was wrong and Pasteur right, Virchow quit science altogether and went into politics. How much wronger could Stubborn Authority get?

But this is a very exceptional case. Let's consider a far more normal and natural example of authority in the wrong.

The example concerns a young Swedish chemical student Svante August Arrhenius, who was working for his Ph.D. in the University of Uppsala in the 1880s. He was interested in the freezing points of solutions because certain odd points arose in that connection.

If sucrose (ordinary table sugar) is dissolved in water, the freezing point of the solution is somewhat lower than is that of pure water. Dissolve more sucrose and the freezing point lowers further. You can calculate how many molecules of sucrose must be dissolved per cubic centimeter of water in order to bring about a certain drop in freezing point. It turns out that this same number of molecules of glucose (grape sugar) and of many other soluble substances will bring about the same drop. It doesn't matter that a molecule of sucrose is twice as large as a molecule of glucose. What counts is the number of molecules and not their size.

But if sodium chloride (table salt) is dissolved in water, the freezing-point drop per molecule is twice as great as normal. And this goes for certain other substances too. For instance, barium chloride, when dissolved, will bring about a freezing-point drop that is three times normal.

Arrhenius wondered if this meant that when sodium chloride was dissolved, each of its molecules broke into two portions, thus creating twice as many particles as there were molecules and therefore a doubled freezing-point drop. And barium chloride might break up into three particles per molecule. Since the sodium chloride molecule is composed of a sodium atom and a chlorine atom and since the barium chloride molecule is composed of a barium atom and two chlorine atoms, the logical next step was to suppose that these particular molecules broke up into individual atoms.

Then, too, there was another interesting fact. Those substances like sucrose and glucose which gave a normal freezing-point drop did not conduct an electric current in solution. Those, like sodium chloride and barium chloride, which showed abnormally high freezing-point drops, *did* do so.

Arrhenius wondered if the atoms, into which molecules broke up on solution, might not carry positive and negative electric charges. If the sodium atom carried a positive charge for instance, it would be attracted to the negative electrode. If the chlorine atom carried a negative charge, it would be attracted to the positive electrode. Each would wander off in its own direction and the net result would be that such a solution would conduct an electric current. For these charged and wandering atoms, Arrhenius adopted Faraday's name "ions" from a Greek word meaning "wanderer."

Furthermore, a charged atom, or ion, would not have the properties of an uncharged atom. A charged chlorine atom would not be a gas that would bubble out of solution. A charged sodium atom would not react with water to form hydrogen. It was for that reason that common salt (sodium chloride) did not show the properties of either sodium metal or chlorine gas, though it was made of those two elements.

In 1884 Arrhenius, then twenty-five, prepared his theories in the form of a thesis and presented it as part of his doctoral dissertation. The examining professors sat in rigid disapproval. No one had ever heard of electrically charged atoms, it was against all scientific belief of the time, and they turned on their built-in doubters.

However, Arrhenius argued his case so clearly and, on the single assumption of the dissolution of molecules into charged atoms, managed to explain so much so neatly, that the professors' built-in doubters did not quite reach the intensity required to flunk the young man. Instead, they passed him—with the lowest possible passing grade.

But then, ten years later, the negatively charged electron was discovered and the atom was found to be not the indivisible thing it had been considered but a complex assemblage of still smaller particles. Suddenly the notion of ions as charged atoms made sense. If an atom lost an electron or two, it was left with a positive charge; if it gained them, it had a negative charge.

Then, the decade following, the Nobel Prizes were set up and in 1903 the Nobel Prize in Chemistry was awarded to Arrhenius for that same thesis which, nineteen years earlier, had barely squeaked him through for a Ph.D.

Were the professors wrong? Looking back, we can see they were. But in 1884 they were *not* wrong. They did exactly the right thing and they served science well. Every professor must listen to and appraise dozens of new ideas every year. He must greet each with the gradation of doubt his experience and training tells him the idea is worth.

Arrhenius's notion met with just the proper gradation of doubt. It was radical enough to be held at arm's length. However, it seemed to have just enough possible merit to be worth some recognition. The professors *did* give him his Ph.D. after all. And other scientists of the time paid attention to it and thought about it. A very great one, Ostwald, thought enough of it to offer Arrhenius a good job.

Then, when the appropriate evidence turned up, doubt receded to minimal values and Arrhenius was greatly honored.

What better could you expect? Ought the professors to have fallen all over Arrhenius and his new theory on the spot? And if so, why shouldn't they also have fallen all over forty-nine other new theories presented that year, no one of which might have seemed much more unlikely than Arrhenius's and some of which may even have appeared less unlikely?

It would have taken *longer* for the ionic theory to have become established if overcredulity on the part of scientists had led them into fifty blind alleys. How many scientists would have been left to investigate Arrhenius's notions?

Scientific manpower is too limited to investigate everything that occurs to everybody, and always will be too limited. The advance of science depends on scientists in general being kept firmly in the direction of maximum possible return. And the only device that will keep them turned in that direction is doubt; doubt arising from a good, healthy and active built-in doubter.

But, you might say, this misses the point. Can't one pick and choose and isolate the brilliant from the imbecilic, accepting the first at once and wholeheartedly, and rejecting the rest completely? Would not such a course have saved ten years on ions without losing time on other notions?

Sure, if it could be done, but it can't. The godlike power to tell the good from the bad, the useful from the useless, the true from the false, instantly and *in toto* belongs to gods and not to men.

Let me cite you Galileo as an example; Galileo, who was one of the greatest scientific geniuses of all time, who invented modern science in fact, and who certainly experienced persecution and authoritarian enmity.

Surely, Galileo, of all people, was smart enough to know a good idea when he saw it, and revolutionary enough not to be deterred by its being radical.

Well, let's see. In 1632 Galileo published the crowning work of his career, *Dialogue on the Two Principal Systems of the World* which was the very book that got him into real trouble before the Inquisition. It dealt, as the title indicates, with the two principal systems; that of Ptolemy, which had the earth at the center of the universe with the planets, sun and moon going about it in complicated systems of circles within circles; and that of Copernicus which had the sun at the center and the planets, earth, and moon going about *it* in complicated systems of circles within circles.

Galileo did not as much as mention a *third* system, that of Kepler, which had the sun at the center but abandoned all the circles-within-circles jazz. Instead, he had the various planets traveling about the sun in ellipses, with the sun at one focus of the ellipse. It was Kepler's system that was correct and, in fact, Kepler's system has not been changed in all the time that has elapsed since. Why, then, did Galileo ignore it completely?

Was it that Kepler had not yet devised it? No, indeed. Kepler's views on that matter were published in 1609, twenty-seven years before Galileo's book.

Was it that Galileo had happened not to hear of it? Nonsense. Galileo and Kepler were in steady correspondence and were friends. When Galileo built some spare telescopes, he sent one to Kepler. When Kepler had ideas, he wrote about them to Galileo.

The trouble was that Kepler was still bound up with the mystical notions of the Middle Ages. He cast horoscopes for famous men, for a fee, and worked seriously and hard on astrology. He also spent time working out the exact notes formed by the various planets in creating the "music of the spheres" and pointed out that Earth's notes were mi, fa, mi, standing for misery, famine, and misery. He also devised a theory accounting for the relative distances of the planets from the Sun by nesting the five regular solids one within another and making deductions therefrom.

Galileo, who must have heard of all this, and who had nothing of the mystic about himself, could only conclude that Kepler, though a nice guy and a bright fellow and a pleasant correspondent, was a complete nut. I am sure that Galileo heard all about the elliptical orbits and, considering the source, shrugged it off.

Well, Kepler was indeed a nut, but he happened to be luminously right on occasion, too, and Galileo, of all people, couldn't pick the diamond out from among the pebbles.

Shall we sneer at Galileo for that?

Or should we rather be thankful that Galileo didn't interest himself in the ellipses *and* in astrology *and* in the nesting of regular solids *and* in the music of the spheres. Might not credulity have led him into wasting his talents, to the great loss of all succeeding generations?

No, no, until some supernatural force comes to our aid and tells men what is right and what wrong, men must blunder along as best they can, and only the built-in doubter of the trained scientist can offer a refuge of safety.

The very mechanism of scientific procedure, built up slowly over the years, is designed to encourage doubt and to place obstacles in the way of new ideas. No person receives credit for a new idea unless he publishes it for all the world to see and criticize. It is further considered advisable to announce ideas in papers read to colleagues at public gatherings that they might blast the speaker down face to face.

Even after announcement or publication, no observation can be accepted until it has been confirmed by an independent observer, and no theory is considered more than, at best, an interesting speculation until it is backed by experimental evidence that has been independently confirmed and that has withstood the rigid doubts of others in the field.

All this is nothing more than the setting up of a system of "natural selection" designed to winnow the fit from the unfit in the realm of ideas, in manner analogous to the concept of Darwinian evolution. The process may be painful and tedious, as evolution itself is; but in the long run it gets results, as evolution itself does. What's more, I don't see that there can be any substitute.

Now let me make a second point. The intensity to which the built-in doubter is activated is also governed by the extent to which a new observation fits into the organized structure of science. If it fits well, doubt can be small; if it fits poorly, doubt can be intensive; if it threatens to overturn the structure completely, doubt is, and should be, nearly insuperable.

The reason for this is that now, three hundred fifty years after Galileo founded experimental science, the structure that has been reared, bit by bit, by a dozen generations of scientists is so firm that its complete overturning has reached the vanishing point of unlikelihood.

Nor need you point to relativity as an example of a revolution that overturned science. Einstein did not overturn the structure, he merely extended, elaborated, and improved it. Einstein did not prove Newton wrong, but merely incomplete. Einstein's world system contains Newton's as a special case and one which works if the volume of space considered is not too large and if velocities involved are not too great.

In fact, I should say that since Kepler's time in astronomy, since Galileo's time in physics, since Lavoisier's time in chemistry, and since Darwin's time in biology no discovery or theory, however revolutionary it has seemed, has actually overturned the structure of science or any major branch of it. The structure has merely been improved and refined.

The effect is similar to the paving of a road, and its broadening and the addition of clover-leaf intersections, and the installation of radar to combat speeding. None of this, please notice, is the equivalent of abandoning the road and building another in a completely new direction.

But let's consider a few concrete examples drawn from contemporary life. A team of Columbia University geologists have been exploring the configuration of the ocean bottom for years. Now they find that the mid-Atlantic ridge (a chain of mountains, running down the length of the Atlantic) has a rift in the center, a deep chasm or crack. What's more, this rift circles around Africa, sends an offshoot up into the Indian Ocean and across eastern Africa, and heads up the Pacific, skimming the California coast as it does so. It is like a big crack encircling the earth.

The observation itself can be accepted. Those involved were trained and experienced specialists and confirmation is ample.

But why the rift? Recently one of the geologists, Bruce Heezen, suggested that the crack may be due to the expansion of the earth.

This is certainly one possibility. If the interior were slowly expanding, the thin crust would give and crack like an eggshell.

But why should Earth's interior expand? To do so it would have to take up a looser arrangement, become less dense; the atoms would have to spread out a bit.

Heezen suggests that one way in which all this might happen is that the gravitational force of the Earth was very slowly weakening with time. The central pressures would therefore ease up and the compressed atoms of the interior would slowly spread out.

But why should Earth's gravity decrease, unless the force of gravitation everywhere were slowly decreasing with time? Now this deserves a lot of doubt, because there is nothing in the structure of science to suggest that the force of gravitation must decrease with time. However, it is also true that there is nothing in the structure of science to suggest that the force of gravitation might *not* decrease with time.[1]

[1] As a matter of fact, there have been cosmological speculations (though not, in my opinion, very convincing ones) which involve a steady and very slow decrease in the gravitational constant; and there is also Kapp's theory, which I described earlier in the book, which involves decreasing gravitational force on earth, without involving the gravitational constant.

Or take another case. I have recently seen a news clipping concerning an eighth-grader in South Carolina who grew four sets of bean plants under glass jars. One set remained there always, subjected to silence. The other three had their jars removed one hour a day in order that they might be exposed to noise; in one case to jazz, in another to serious music, and in a third to the raucous noises of sports-car engines. The only set of plants that grew vigorously were those exposed to the engine noises.

The headline was: *BEANS CAN HEAR—AND THEY PREFER AUTO RACING NOISE TO MUSIC.*

Automatically, my built-in doubter moves into high gear. Can it be possible that the newspaper story is a hoax? This is not impossible. The history of newspaper hoaxes is such that one could be easily convinced that nothing in any newspaper can possibly be believed.

But let's assume the story is accurate. The next question to ask is whether the youngster knew what he was doing? Was he experienced enough to make the nature of the noise the only variable? Was there a difference in the soil or in the water supply or in some small matter, which he disregarded through inexperience?

Finally, even if the validity of the experiment is accepted, what does it really prove? To the headline writer and undoubtedly to almost everybody who reads the article, it will prove that plants can hear; and that they have preferences and will refuse to grow if they feel lonely and neglected.

This is so far against the current structure of science that my built-in doubter clicks it right off and stamps it: IGNORE. Now what is an alternative explanation that fits in reasonably well with the structure of science? Sound is not just something to hear; it is a form of vibration. Can it be that sound vibrations stir up tiny soil particles making it easier for plants to absorb water, or putting more ions within reach by improving diffusion? May the natural noise that surrounds plants act in this fashion to promote growth? And may the engine noises have worked best on a one-hour-per-day basis because they were the loudest and produced the most vibration?

Any scientist (or eighth-grader) who feels called on to experiment further, ought to try vibrations that do not produce audible sound; ultrasonic vibrations, mechanical vibrations and so on. Or he might also try to expose the plant itself to vibrations of all sorts while leaving the soil insulated; and vice versa.

Which finally brings me to flying saucers and spiritualism and the like. The questions I ask myself are: What is the nature of the authorities promulgating these and other viewpoints of this sort? and How well do such observations and theories fit in with the established structure of science?

My answers are, respectively, Very poor and Very poorly.

Which leaves me completely unrepentant as far as my double role in life is concerned. If I get a good idea involving flying saucers and am in the mood to write some science fiction, I will gladly and with delight write a flying-saucer story.

And I will continue to disbelieve in them firmly in real life.

And if that be schizophrenia, make the most of it.

17 Battle of the Eggheads

After the Soviet Union placed Sputnik I into orbit on October 4, 1957, the egghead (to use a term invented by a blockhead) gained a sudden, unaccustomed respect here in the United States. Suddenly everyone was viewing American anti-intellectualism with wild alarm.

It has therefore always tickled my vanity that I wrote an article deploring anti-intellectualism in America a year and a half *before* Sputnik.[1]

[1] "The By-Product cf Science-Fiction," *Chemical and Engineering News*, August 13, 1956.

In it, I disapproved vehemently of those factors in American culture which seemed to me to be equating lack of education with virtue and to be making it difficult for young people to reveal intelligence without finding themselves penalized for it.

I said all this without mentioning missiles or satellites, without any talk of a "scientific race" with any nation. In fact, I never mentioned the Soviet Union at all. As I said, this was one and a half years before Sputnik I, and before the flood of Monday-morning quarterbacks, wise after the event, that followed hard upon Sputnik I's launching.

Of course, I must hastily disavow any intention of trying to imply that I'm smarter or more prescient than the next fellow. I did not foresee Sputnik I. An astronomer I know warned me in the spring of 1957 that the Soviet Union might beat us to the punch and I laughed heartily and confidently. "Never," I said.

But that only means I never thought intelligence was important just because we had to keep ahead of the Soviet Union. I thought intelligence was important for various other good and sufficient reasons, and sounded the trumpets on its behalf even when I was convinced that the United States was safely ahead of all comers in all branches of science.

So after I recovered from my amazement that October day, I sat back to marvel at the sudden prestige that brains fell heir to; and to wonder at the spectacle of congressmen discussing spaceflight learnedly, just as if they had been reading up on science ever since they kissed their first baby. For a while, it seemed to me that brains had grown so respectable that I thought I could detect congressmen trying to speak grammatically, even though that meant losing their All-American flavor of rough-hewn backwoods virtue.

In those days everyone talked about revising our system of education, and introducing the revolutionary system of actually encouraging the brighter schoolboys and paying them some attention.

But then, initial panic subsided. We sent up a number of satellites of our own and "Yankee know-how" was a phrase to conjure with again. That left room for the thought that, after all, better schools cost money and who can afford to throw money away by paying schoolteachers full-scale janitorial-type salaries?

What's more, something else was added. Complacency and false economy are nothing over which to be shocked, for anyone who is surprised by the existence of either had better turn in his sense of cynicism for a sharper-edged model.

The "something else" to which I refer (and which *is* shocking) is a definite counterattack against any changes in our basic educational philosophy and against the whole notion of increasing emphasis on science on the part of some of the eggheads themselves.

After all, there are eggheads and eggheads, in a variety of genera and species. We can make a broad classification, however, and divide them up into the humanists and the scientists (which doesn't mean, of course, that one man can't be a member of both groups).

There is snobbery among the educated; there always has been. As long ago as the time of ancient Greece, the great philosophers felt quite certain that to investigate nature by deep and abstract thought was far superior to, and nobler than, investigation by experimentation. They felt that to delight in the beauty of the ordered universe out of a pure appreciation of the aesthetic, was superior to an interest grounded in a desire to apply the laws of the universe to the uses of everyday living.

Perhaps this was because Greece was a society founded on human slavery, so that there grew to be something disgraceful about manual labor. Experimentation, after all, was a kind of manual labor and therefore fit only for slaves, really. Applied science meant bending the glories of the universe to those things that should interest slaves. The very expression "liberal arts" comes from the Latin *liberi* meaning "free men." The liberal arts were suitable for free men; the mechanical and technical arts for slaves.

A great thinker such as Archimedes, who couldn't resist working in applied science (and doing it superlatively well, too), was nevertheless ashamed of himself and would publish only his theoretical work.

So experimental science had to wait two thousand years to be born.

And the attitude persists today, even among the experimental scientists themselves. The more theoretical a science, the higher it is in the scientists' social scale. The descending hierarchy of science is: mathematics, astronomy, physics, chemistry, biology, and sociology. Within each discipline, there are subdivisions that can be similarly treated on the basis of theoretical content. Within chemistry, for instance, the descending hierarchy is: physical chemist, organic chemist, biochemist, chemical engineer.

It is interesting that the various major disciplines of science developed their modern contents in the order of their position in the hierarchy, as though it took longer and longer for thinkers to break further and further from the Greek ideal.

Modern sociology did not really come into its own until the twentieth century (and perhaps hasn't, even yet, really got off the ground). Modern biology—including the cell theory, the germ theory of disease, and the theory of evolution by natural selection—is a nineteenth-century creation. Modern chemistry is the creature of Lavoisier and the eighteenth century; modern physics of Galileo and the seventeenth century. Modern astronomy dates back to Copernicus and the sixteenth century.

Mathematics, finally, is so highly theoretical that the Greeks condescended to invent it in the modern sense. Furthermore, it never entirely died in the centuries of darkness after. By the fifteenth century, mathematics began to show unmistakable signs of a renewed vitality that has never flagged since.

But what lies beyond mathematics and the fifteenth century? What most-high of modern life came into being in the fourteenth century? Answer: the humanities.

Practitioners of all the sciences alike feel themselves (consciously or not) to be culturally inferior to those who specialize in the humanities. The humanists balance this situation by feeling smugly superior to the scientists, and because in the very nature of the case, the humanists are extremely articulate, they have sold this attitude to the public generally.

When any of us think of culture, we think of literature, art, music, philosophy, Latin, Greek—things like that. And in fact, so untouchable have "things like that" become, that in beginning a discussion intended to be iconoclastic about them, I almost feel as though I were going to denounce mother love or refuse to salute the flag or something equally heinous.

Now what are the "humanities" anyway? Webster says: "The branches of polite learning regarded as primarily conducive to culture: especially the ancient classics and belles-lettres; sometimes, secular, as distinguished from theological, learning."

The first part of the definition makes it seem obvious that the humanities are a type of "pure" learning not readily applied to the everyday problem of making a living. It is an ideal study for leisure time and for those people who have leisure time.

And it is only human to fall into the fallacy that if *a* implies *b*, then *b* must imply *a*. If the best examples of the humanities have no practical application, then studies without practical application are good examples of the humanities; and, conversely, a study *with* a practical application is *not* a good example of the humanities; it is not a type of polite learning, it is not conducive to culture.

Now the various sciences can't avoid having practical uses. The sciences start with gentlemen amateurs but invariably end with someone in a laboratory somewhere getting himself all dirty.

Who would therefore argue that the immensely learned gentleman with the vast world of the humanities at his fingertips, but with no knowledge of science was not far more cultured than the laboratory worker with detailed knowledge of the sciences but unable to differentiate between a Picasso and a pizzicato.

For instance, there is a story that the faculty of the Massachusetts Institute of Technology once met to go over the final grades of the graduating class. It turned out that a student named Cicero had flunked Latin. There was general laughter in which all, without exception, joined.

Who did not know, whatever his specialty, that Marcus Tullius Cicero was the greatest of the Roman orators and the writer of the purest specimens of Latin style ever committed to paper? Not to know that was to be uneducated and boorish. The physicist, as well as the classicist, would have been ashamed not to know.

There then followed, at this same faculty meeting, the case of another student, named Gauss, who had flunked mathematics. Again there was laughter, but now only the members of the various science departments laughed. The humanities boys maintained an uncomprehending silence.

They did not know that Karl Friedrich Gauss was one of the three or four greatest mathematicians of history. And if that were explained to them, they undoubtedly didn't see why they should be expected to know and probably didn't care that they didn't know, and had every intention of not knowing the next time, either.

After all, any scientist would be ashamed to look up from his instruments and say, "I don't dig this fancy literature jazz. I just read comic books." It might be true, mind you, but he would be ashamed to say so. He would feel disgraced.

However, I can easily imagine a humanist stating quite calmly that he knew nothing about mathematics and that he couldn't add a column of figures to save his life. There's no disgrace in *that*. In fact, I have a suspicion that a thoroughgoing humanist would feel just a little proud of not understanding mathematics or science. It would be a sign of true intellectual aristocracy. It would show how *completely* cultured he was.

So now consider the situation in which the humanists found themselves unexpectedly involved after that black October day in 1957. The American public and its spokesmen were suddenly howling for more education, but it was *science* education they were speaking of. Eminent leaders in all walks of life suddenly discovered that our youngsters weren't being taught enough—*science*, that is.

Imagine the possible future that faced the thoroughly cultivated humanist. Would the time come when a man was to be considered educated simply because he understood differential equations, Lord save us? Was a chemist, with his acid-stained fingers and his stinks, to be looked up to as a cultured individual, *ipso facto?*

And what would happen to a man, a *really* cultured man, who had read Proust in the original French and Dostoevski in the original Russian (Czarist Russian, of course), but who had never quite sullied himself with calculus and protons and things like that. Was he to be a mere layman? Was he to be a person with a second-class education?

Naturally, many humanists were against any such development. The fellow with the buggy-whip factory was as naturally against Henry Ford.

The result has been a learned counterattack against any "overemphasis" on science for a variety of reasons, some of which I find more sickening than others.

One point I hear often made is that we are allowing the Soviet Union's successes in the missile field to drive us into a base competition with an evil materialistic society in turning out scientists and engineers; that we should instead follow our own more spiritual way of life; that we should not try to defeat a hellish system by adopting the vile features of the very thing we are fighting.

Of course, there is to me something foolish and hypocritical about making it look as though we were too proud to compete with the Soviet Union on a material basis. It wasn't many years ago when it was loudly stated that all we had to do was to drop Sears, Roebuck catalogs all over the Soviet Union; that the suppressed population, learning of the vast wealth and riches made possible by an enlightened capitalistic system, would then rise in rebellion.

We proved our superiority over communism again and again, by the simple process of comparing numbers of automobiles, telephones, washing machines, and such. Anyone watching television knows that our economy is dependent on continually increasing the count of all our material possessions and that all extremes are taken to encourage this. If some method were discovered that would enable an announcer to emerge from the set and shove his detergent or headache powder or cake mix or automobile down our throats at the point of a gun, sponsors would undoubtedly stand in a double line three blocks long waiting their turn.

Now, after forty years of listening to this, the Soviet Union suddenly brightens up and says, "All right, we'll beat you out in standard of living and let's begin right now by measuring who is superior in terms of numbers of missiles and scientists." If, in response, all we can do is mutter that oh, well, it's the spiritual values that count after all, I can only say that this conversion is too late to be convincing, and we're going to lose that celebrated fight for the minds of men.

Then too, I am horrified at any grisly line of reasoning that tends to make it look as if educating more scientists is somehow to adopt a communist line. To equate science with communism is plain suicide for any non-communist society. And frankly, if to be in favor of more and better science and scientists is to be communistic, I might as well turn myself in right now.

What's more, to talk about "competing with the Soviet Union" is to miss the point anyway; to miss it so badly that I stand appalled at the naïveté of it all. Suppose the Soviet Union were to be converted to Jeffersonian democracy tomorrow, or to pure Christianity; or better yet (since we wouldn't trust them anyhow, I imagine) suppose the Soviet Union with its entire population were to vanish this instant from the face of the Earth.

Do you suppose there would then no longer be any need for science or scientists; that we could all sit back and listen to a Brahms concerto or an Elvis Presley record (according to taste) and leave science to the few queers and oddballs who have an ingrained, unstoppable interest in it?

Not on your set of the great works of literature of all ages. We could not.

We have an enemy worth ten times the Soviet Union called "exploding population." We've got another one, equally formidable, called "declining resources." And we've got a world population who have either attained a high standard of living and want more of the same, or who have not yet attained a high standard of living and are determined to do so.

I don't have to beat these particular dead horses in detail, but I wish to point out just briefly that if we expect to make an easier life for more people out of what's left of a plundered planet, we're going to have to look for ways of doing it. Belles-lettres may inspire us in this search, but the actual answers, if any, will have to come out of the scientific disciplines.

We'll need scientists and engineers for more than missiles and satellites. We'll need them for such things, such simple everyday things, as finding food, pure water, and uncontaminated air.

In fact, we'll need scientists to a greater degree if the Soviet Union disappears; because while the Soviet Union exists, there's always the chance that the whole works may be queered by means of a full-scale nuclear war and then we won't need science at all for a while—or much of anything else, either.

To be sure, it can be pointed out that many modern problems would not exist were it not for science. The danger of nuclear war is the best example. Then too, the advance of modern medicine has been one factor behind the present population explosion; and, as an example of a minor item, it appears to be automobile exhaust that creates the comedian's delight and the lung-owner's despair—smog.

However, science did not invent the problem of having problems. Problems existed in plenty in the days of non-science and made non-scientific societies far more miserable than our own in many respects and with far less hope of relief. The ideal civilization of the humanist, Periclean Athens, was founded on human slavery and lasted one generation, being then destroyed by war (that was chronic in those times and certainly not caused by science) and plague (that was also chronic and *was* caused by non-science).

I think that anyone who yearns for some simpler pastoral society, some primitive and virtuous patriarchal culture away from the madness of modern life, yearns for something that never was.

It may be crassly materialist of me but I get a warm feeling of comfort and security when I think of such things as anesthetics and antibiotics and soap and internal plumbing and a million other things that Daphnis and Chloë did not have as they piped interminable pipings to their gamboling lambs. And what do you suppose happened to Daphnis anyway if he came down with an attack of acute appendicitis? He didn't scream, forever, of course; only till he went into a coma and died.

Another fear frequently expressed in connection with possible "overemphasis" on science, is that we might turn out a nation of scientific robots; that it is important, after all, to turn out "well-rounded" men.

This is ignorance, which is bad; or hypocrisy, which is worse. It consists of raising the specter of a horrible danger that does not and cannot exist. Let's suppose that the American people *want* to turn out a nation of scientific robots; that American education has accepted the challenge and goes honestly to work toward that goal. Will it succeed? Of course not.

The large majority of the human race are no more equipped to be brilliant scientists than they are equipped to be star football players. With the best will in the world we could only turn a minority of even the gifted portion of humanity into high-class probers into the secrets of Nature.

The term "scientific robot," which is used frequently by humanists, is an unjustified piece of intellectual snobbery, in which the humanist joins with the generally uneducated in accepting a false stereotype of the scientist, as someone who is lost in his test tubes and oscilloscopes and is incapable of appreciating the finer things of life.

Despite a wide acquaintance in the field, I know very few scientists who are lost in either test tubes or oscilloscopes. Most have outside interests; among other things, in the humanities. Most believe that a man is a better scientist for being interested in the humanities and act on that belief. As it happens, I *do* know someone who has read Proust in the original French and Dostoevski in the original Russian. He's a biochemist.

Still another great worry about concentrated science-teaching is this: Suppose you *do* decide on a great and continuing drive to find and develop students who are capable of scientific work. Suppose you concentrate on making those students scientists. Are you not then destroying the student's right to lead his own life, choose his own interests? Suppose a student who can be a scientist doesn't wish to be a scientist? Is not making him one anyway anti-democratic? Is it not dictatorial? Is it not interfering with the human dignity and individuality that the Western world has struggled so hard to preserve?

The answer is, yes, to every one of those questions, and if a student is dead set against being a scientist, he can't be made into one, no matter how qualified he is otherwise. The only thing is that we had better make sure that he *is* dead set against being one. We had better give him every inducement first to be a scientist.

I remember the naïve, dreamy days before Pearl Harbor when there arose the question of establishing a draft. Some great minds in the halls of Congress rose to say that a draft was unnecessary because at the first hint of invasion, a million Americans, like the good old minutemen of old, would spring to arms.

Sure, they would!

The minutemen of old grabbed their long squirrel rifles off the wall and went out to shoot at redcoats who didn't have any gun they could handle half as well. And the American of 1941, presumably, would grab his tank and airplane off the wall and go running.

Fortunately, by the spanking majority of one, our noble leaders in the House of Representatives dimly made out the fact that modern weapons can't be handled at first sight; that there's more to war these days than pulling a trigger. So the draft went into action and when war came we only needed an additional six months to get ready.

Now the army draft is dictatorial and destroys individuality. No one asks the recruit if he would rather be a private or a rich war worker. Alas, necessity rules after all.

Actually, we are at war now; not with the Soviet Union, but with the universe. We always have been. Human progress—or what we are pleased to call progress—came about as the result of victories over the universe. There was the discovery of fire, the invention of the wheel, the development of metallurgy, the taming of the horse.

After 1500, an organized method of fighting the universe was invented and called experimental science. After 1750, that method went into high gear. Until 1950, however, the war against the universe was still small-scale enough to be carried out with reasonable efficiency by a volunteer army.

That's no longer the case. Increasing population plus the intense scale of energy expenditure made possible by earlier victories has made the battle continually more involved and the risk of disaster in case of defeat (even temporary defeat) continually higher.

A volunteer army is no longer enough. We need a draft in the form of a revised and improved educational system; and the assurance that every man who is equipped to be a scientist, both intellectually and psychologically, become one. We need to make sure that no budding scientist be lost to humanity for trivial reasons.

To put it bluntly, I would also like to see an end to dilettantism in matters of the intellect. The Greek exaltation of art for art's sake is fine as long as it isn't interpreted to mean that art for the good of mankind is ignoble.

I say, let's consider the second part of the Webster definition of the humanities, which reads: "—secular, as distinguished from theological, learning."

The humanities, in the modern sense, were invented during the Renaissance. At that time, when European education had long been centered about theology, Italian scholars rediscovered the secular literature of Greece and Rome; a literature that concerned itself not with Heaven and Hell but with the things of this earth. The ancients had a view of life that dealt with man and his relations with man; and this was ravishingly novel to a culture that had concerned itself for a thousand years with God and his relation to man.

So the scholars of the Renaissance became concerned with "humanity," rather than with "divinity," and what they studied was called "the humanities" in consequence. The humanities had to concern themselves first with "the ancient classics" (as Webster says) and the imitative literature of "belles-lettres," which the humanists themselves wrote.

But because that is the way it started doesn't mean that that is the way it must end.

The humanities are secular learning; they are the study of that which concerns men, and in the centuries that have progressed since the Renaissance, new things have come to concern men. Are the new things to be forever unrepresented? Modern science is a creature of the post-Renaissance, and because Francesco Petrarca knew nothing of it, is that a reason we must know nothing of it, either?

In the modern world, science plays an intimate role in all aspects of man's life. From top to bottom, from mind to belly, we live surrounded by and permeated with science and the objects that are the products of science. It is impossible any longer to divorce man from science or science from man, without unimaginable catastrophe.

Therefore the man who calls himself a humanist but remains ignorant of science is not really a humanist, because he has, more or less deliberately, cut himself off from one of modern humanity's most important concerns.

This does not mean a humanist must today be a professional scientist. Of course not. No one expects him to be a great novelist or to compose a sonata or to turn out a still-life sketch. He is expected, however, to know something about literature, music, and art, and to appreciate them. He should also be expected to understand something about science and appreciate that.

If that attitude came to pass, we could develop a new group of twentieth-century humanists, men who could forsake *quattrocento* outlooks and prejudices and join the rest of us way up here in the present. With his new outlook, the humanist may then not be so unreasonably frightened at our modern need to intensify science education and perhaps then, marching forward under the emblem "Eggheads United," we can continue to gain in the never-ending battle against the Universe.